Wayward Technology

Wayward Technology

Ernst Braun

Contributions in Sociology, Number 48

GREENWOOD PRESS
Westport, Connecticut

Published in the United States and Canada by
Greenwood Press, a division of Congressional
Information Service, Inc., Westport, Connecticut

First published 1984

Library of Congress Cataloging in Publication Data

Braun, Ernst
  Wayward technology.

  (Contributions in sociology, ISSN 0084–9278   No. 48)
  Includes index.
  1. Technology—Social aspects   I. Title
II. Series
T14.5.B73   1984   600   83–22586
ISBN 0–313–24398–0   (Lib. Bdg.)

Library of Congress Catalog Card Number: 83-22586

ISBN: 0-313-24398-0

Printed in the United States of America

Typeset by Joshua Associates, Oxford
Printed in the USA

The spirits I have called
I cannot now dispel

From 'The Sorcerer's
Apprentice' by
J. W. Goethe

Contents

List of Figures

List of Tables

Preface

Technology is a many-faceted wayward creature. It is of society, yet much of its activity seems to be concerned but with itself. This book is an attempt to illuminate some facets of this social activity called technology. Such illumination should be its own reward, yet at the back of my mind there is the sneaking hope that inspection may yield some guidance to possible ways of increasing the social utility of technology.

The plan of the book is straightforward. The first chapter sets the scene in describing some of the main features of technological society and showing by what circuitous paths technology came to occupy the central stage of society. The next chapter takes us into the present and describes how new products of technology emerge. The third and fourth chapters deal with contemporary fears about the apparently uncontrolled and dangerous paths technology seems to take and with one proposed solution to these problems—technology assessment. Chapters 5 and 6 discuss public and private forces which influence the course of technology, while Chapter 7 asks what society might want from its technology. The last chapter is a brief summary of the elemental forces which influence technology, a glimpse at the likely future faces of technological society and a cursory look at avenues which technology policy might pursue in order to obtain a more desirable future society.

This book would not have been written without the help of many of my friends. First and foremost among them are my former colleagues, past and present staff and students of the Technology Policy Unit of the University of Aston in Birmingham. This really is their book. Many other friends in England, Austria and Germany have helped by discussion, encouragement and critical reading of early drafts. To them all I offer my sincere thanks and gratitude. The book was written during a period of leave of absence from the University of Aston and my thanks are due and gladly offered to my hosts during this period, the Technical University of Vienna during two semesters and the Wissenschaftszentrum Berlin during the summer.

Lynton, Devon, May 1983   Ernst Braun

Wayward Technology

1　The Rise of Industrial Society

Humans are physically ill-endowed for survival in most of the climates in which they live. They owe the fact that they have not only survived but have come to dominate over all other creatures to cooperation and to technology which grew out of it. Only co-operation ensured the provision of the bare necessities of life and cooperation was inextricably bound with division of labour and the use of tools. It is idle to speculate whether tools and techniques are the cause or the result of cooperation, what matters is that since the beginning of mankind humans banded together and used tools and techniques to overcome hostile forces, whether of the elements, animals or other humans. Gradually they learned to provide for themselves comforts beyond those required for sheer survival. This became possible only by more efficient tools and by more efficient means of using them. Technology and social organisation may be two faces of mankind, but they point firmly in the same direction: improved chances of survival; improved living conditions; the creation of surpluses beyond the bare necessities. Some of these surpluses could be used to create social rituals which catered both for spiritual needs and increased social cohesion, thus strengthening the bonds which alone assured survival and further surpluses. *Homo Faber* and *Homo Ludens* are both firmly rooted in the very origins of mankind, and technology has always occupied a central position in human affairs. Yet the conflict between things material and things spiritual, the conflict between the two faces of mankind, is probably equally firmly rooted in human history. Although we have moved a long way from the primitive tools and elementary ritual of our ancestors, the conflict and the interdependence between our material, spiritual and social needs are as potent as ever.

Man became human by cooperation in tasks of survival. Great advantages were obtained by hunting together and equally great or greater advantage by banding together in common defence. Cooperation creates and is a result of social cohesion. People who cooperate with each other develop social structures and a sense of belonging to a social unit. Conversely, people who feel a sense of community do cooperate with each other in the fulfilment of economic tasks.

Homo Faber is thus greatly aided by cooperation—Cooperative Man achieves a great deal more than he can in isolation. Cooperation increases economic efficiency and makes it possible to create surpluses beyond the minimum requirements of mere survival. These surpluses, however small, enable humans to develop the second facet of their nature—the need for ritual, for organised play. The surplus created by cooperation can be used to increase social cohesion by the creation of ritual; and social cohesion increases cooperation. We may say that indeed not only are *Homo Faber* and *Homo Ludens* the two facets of *Homo Sapiens* but we may also identify interdependence between them. *Homo Faber*, the economic cooperator, enables *Homo Ludens* to indulge in play and create social forces for cooperation.[1]

Economic cooperation, cooperation initially in tasks of survival and later more and more in the satisfaction of further needs, created technology. Indeed we may define technology as the creation and use of artefacts for practical purposes. No matter that this definition begs some questions, it shows that the satisfaction of needs—the achievement of *Homo Faber*—is achieved through the use of tools in the broadest sense and these tools are created and used largely in cooperative ventures. The tragic contradiction of modern technology is that it is destroying the very cooperation which created it.

In advanced modern technological society the time occupied by activities essential to simple survival has become a very small fraction of the total time available. Even when we make things now, more often than not these things are not essential for survival. Our shops are filled with fanciful items used for entertainment, sport, fun or keeping up with the Joneses. Even apparently essential items of consumption, such as food, contain unnecessary additives, packaging, over-refinement. The distinctions between *Homo Faber* and *Homo Ludens* have become more blurred than ever.

Technology has long been recognised as one of the major forces shaping society. Indeed we speak of stone age man and iron age man, characterising ancient societies by their primary technology. But not only ancient societies have been thus described; one of the principal features of much modern social analysis is the central position occupied by means and relationships of production in social relationships.[2] The expression 'means of production' is synonymous with 'technology of production' and relationships of production subsume, *inter alia*, work organisation, including the factory system of production.

If the recognition of the central position of technology in society is not new, there is one feature of modern technological society which, although not entirely new, has reached alarming proportions. While in previous eras technology was regarded as pure achievement, with some dissent and an occasional tinge of fear, in our modern era the sense of achievement is almost balanced by a sense of loss of control and a desire to regain what is imagined to have been human dominance over technology in the past. There is now an overwhelming impression that society has come to be at the mercy of its technology, a technology that advances rapidly under its own momentum and irrespective of whether it is socially good or evil.[3]

Yet never before has society made such strenuous efforts not so much to control technology as to help advance it. For despite all fear of technology and all the feeling of being overrun and perhaps destroyed by it, societies feel that technology is their best weapon in gaining competitive advantages over rival societies. Contrary as it may seem, technology is regarded with fear and suspicion and yet is cosseted as Man's best friend in his struggle not only to survive but to get ahead of the pack. Perhaps we still live the Promethean myth—we want the fire yet fear punishment for using it. Perhaps also the immediate economic interest of any given social entity militates against the wider long-term interests of human society at large.

It is a central task of this book to shed some light on this dilemma —the dependence upon technology yet the fear of submission to it. To this end we must attempt to understand the twin question: what forces shape technology, and how can and does society control these forces? But in the very posing of this question we must recognise that setting technology apart, we deny its essential existence as an organic part of society. Thus we set it apart only to examine it, isolate it for purposes of analysis, recognising all the time that technology is of society, indeed the very essence of social activity.

TECHNOLOGICAL SOCIETY

When we speak of contemporary society as 'technological society', we have a very specific meaning in mind and distinguish this society from all others, although in a wider sense all societies are technological, for all are shaped by their technologies. Contemporary technological society is distinguished by a number of primary features. It is a society which has

  (i) a system of manufacture of large quantities of a great variety of goods in a multitude of industrial enterprises, some of considerable size;

 (ii) a complex infrastructure of transport, communication, education, distribution and other services as well as provision for a wide range of consumer services;

(iii) a highly developed system of scientific-technical activities which continually replenish the technology used in providing goods and services and enlarge the range of goods and services available.

The primary features are associated with an extensive set of secondary features, only a few of which shall be listed:

  (i) large conurbations to provide adequate supplies of labour, an adequate network of interdependent service and manufacturing organisations and adequate markets for a wide variety of consumer services. The conurbations do, of course, require their own extensive infrastructure of services, including sanitation, transport, security and retail distribution;

 (ii) an adequate supply of energy and of raw materials;

(iii) adequate supplies of food and other agricultural products, produced by efficient methods requiring only a small proportion of the total workforce, thus leaving sufficient workers for industry and services;

 (iv) a great deal of division of labour and specialisation and hence much interdependence and trade within the manufacturing system and its services;

  (v) last, but certainly not least, industrial production can only be established and maintained if there are means of accumulating economic surpluses and converting these into productive machinery. In other words, the accumulation of capital and its investment into industrial equipment are pre-requisites for any industrial operation. A mere subsistence economy cannot industrialise.

The industrial system of a modern developed country is vast and the total production (and consumption) immense. A few tables describing some aspects of the British economy in the 1970s may help to illustrate the point (see Tables 1.1 to 1.5).

Technological society is derived from earlier industrial society, indeed the difference lies only in the extensive and systematic use of science by technological society and in the quantity of production. Whereas industrial society relied more upon craft-based

Table 1.1. United Kingdom Gross National Product by origin, 1974

	£m	% of GDP
Agriculture, forestry and fishing	2,116	2.9
Mining, quarrying	1,021	1.4
Manufacturing:		
Food, drink, tobacco	2,410	
Chemical and oil products	2,012	
Metal manufacturing	1,271	
Engineering and allied industries	8,981	
Shipbuilding and marine engineering	466	
Vehicles and aircraft	1,883	
Textiles	1,119	
Clothing	639	
Other manufacturing	4,213	
Total manufacturing	22,994	31.7
Construction	5,645	
Gas, electricity, water	2,255	
Transport and communication	6,648	
Distributive trades	7,003	
Insurance, banking, finance / Professional and business services	6,750	
Miscellaneous services	8,735	
Total services	37,036	51.0
Public services and administration, defence, ownership of dwellings, adjustments for financial services and residual error	9,458	13.0
Total GNP	72,625	

Source: After Prest, A. R., and Coppock, D. J. (eds), *The UK Economy*, London, Weidenfeld & Nicolson, sixth edition, 1976, p. 188.

industrial processes, technological society relies upon a highly organised system of research and deliberate technological innovation, coupled with an extensive and complex information gathering and processing system. To understand any particular era, it is necessary to go back to its identifiable origins. But technology is as old as mankind—ignoring for our purposes the prelude in the Garden of Eden—and we must choose a suitable beginning other than the fall from grace. The most convenient point of departure is the industrial revolution in Britain, for it created the first industrial society.

Table 1.2. Size distribution of manufacturing units by employment, United Kingdom, 1972

Size category	No. of units	%	No. employed	%
11–19	17,155	28.7	246,722	3.3
20–24	8,150	13.6	179,045	2.4
25–99	20,637	34.5	1,049,668	14.2
100–199	6,246	10.4	867,511	11.7
200–499	4,894	8.2	1,509,389	20.4
500–999	1,589	2.7	1,101,187	14.9
1,000 and over	1,121	1.9	2,433,682	32.9
Total	59,792	100.0	7,387,204	99.8

Source: After Prest and Coppock (eds), *The UK Economy*, p. 230.

Because technological society is deeply rooted in industrial society, our understanding of the former may be deepened by enquiry into the latter. It is helpful to cast a very brief glance at the origins of industrial society during the period of the industrial revolution, as some of the forces causing industrial change today are similar to those discernible during that period.

THE INDUSTRIAL REVOLUTION

The dating of such a complex web of events as those known as the industrial revolution in Britain contains a large degree of arbitrariness and convention. It is difficult enough to date a revolution which erects barricades, let alone one that consists of a long series of many interrelated events which, taken overall, transformed a largely rural society into a largely industrial one over a relatively short period of time.

It is generally agreed that up to the middle of the eighteenth century, say to 1760, British society was predominantly agricultural and rural. This does not mean that there were no cities, nor that there was no manufacture of goods or trade in manufactured articles. Nor does it mean that there was no mining or even that none of the many important mechanical inventions, so characteristic of the industrial revolution, were made before 1760. Indeed even the steam engine goes back to an earlier date. What it does mean is that the agricultural sector was dominant in the economy, that both the creation and ownership of wealth were strongly linked to land and

Table 1.3. Energy consumption: vehicle and steel production, United Kingdom, 1969–80*

	Primary Fuel input (Million tonnes of coal or coal equivalent)				Electricity available (GB) Thousand million kilowatt hours	Passenger cars Monthly averages, thousands	Crude steel production Thousand tonnes
	Total	Coal	Petroleum	Natural Gas			
1969	323.0	162.8	138.4	9.4	200.5	143.1	26,822
1970	336.3	156.8	149.8	17.8	211.6	136.7	27,792
1971	332.9	140.0	152.4	28.8	215.9	145.2	24,153
1972	337.4	121.8	162.2	41.0	224.1	160.1	25,293
1973	354.2	133.2	164.5	44.4	241.0	145.6	26,594
1974	337.9	117.7	153.1	52.9	235.5	127.8	22,323
1975	326.9	120.6	137.1	56.3	234.2	105.6	19,789
1976	331.5	123.0	134.2	59.5	236.7	111.1	22,274
1977	338.2	122.4	136.8	62.7	240.2	109.7	20,411
1978	339.1	119.4	139.2	65.0	246.3	101.9	20,311
1979	349.5	127.2	137.0	69.3	255.0	89.2	21,412
1980	326.5	120.2	121.2	69.9	245.9	77.0	11,387
1975 3	326.4	116.9	139.2	57.4	234.7		
4	318.9	117.2	129.2	58.5	228.5		
1976 1	332.5	125.5	133.8	60.4	230.9		
2	334.6	120.8	139.5	59.2	240.4		
3	330.9	120.5	136.7	57.6	244.3	106.8	5,375
4	328.0	125.2	126.8	61.0	231.4	112.1	5,650

Table 1.3. (cont.)

		Primary Fuel input (Million tonnes of coal or coal equivalent)				Electricity available (GB) Thousand million kilowatt hours	Passenger cars Monthly averages, thousands	Crude steel production Thousand tonnes
		Total	Coal	Petroleum	Natural gas			
1977	1	336.9	126.3	134.0	61.1	237.9	104.6	5,417
	2	337.7	121.1	137.7	62.0	242.9	114.8	4,908
	3	340.5	121.6	138.1	63.4	248.3	108.9	5,419
	4	337.5	120.4	137.3	64.3	231.9	110.3	4,641
1978	1	331.4	116.7	133.4	64.9	237.3	114.9	4,957
	2	341.2	118.7	142.9	64.2	248.1	109.4	5,278
	3	338.3	117.2	143.6	62.4	253.9	105.7	4,757
	4	343.4	125.2	136.7	66.5	245.7	77.6	5,245
1979	1	349.1	121.8	141.0	70.3	252.6	102.7	4,954
	2	355.8	131.3	139.4	69.4	257.6	98.1	5,720
	3	352.1	131.8	137.8	66.6	257.7	72.2	5,485
	4	342.1	124.8	129.8	71.1	252.0	83.8	5,252
1980	1	333.9	121.5	124.4	72.4	242.5	92.4	764
	2	328.3	122.1	122.0	69.9	248.1	75.7	4,098
	3	323.7	119.3	123.1	66.1	249.0	75.4	3,498
	4	321.2	117.6	115.4	71.7	244.1	64.5	3,027

*Annual rates seasonally adjusted and temperature corrected.

Source: Central Statistical Office, Economic Trends. London, 1981, pp. 24 and 32 (with permission of the Controller of Her Majesty's Stationery Office).

Table 1.4. United Kingdom inland freight transport, 1957–74
('000m ton-miles)

Year	Road	Rail	Coastal shipping	Inland waterways	Pipelines*	Total
1957	22.9 (42.5%)	20.9 (38.8%)	9.8 (18.2%)	0.2 (0.4%)	0.1 (0.2%)	53.9
1965	42.1 (57.1%)	15.4 (20.9%)	15.3 (20.8%)	0.1 (0.1%)	0.8 (1.1%)	73.7
1974	53.0 (65.1%)	14.3 (17.6%)	11.9 (14.6%)	0.1 (0.1%)	2.1 (2.6%)	81.4

*Excludes movement of gases.
Source: After Prest and Coppock (eds), *The UK Economy*, p. 198.

Table 1.5. United Kingdom inland passenger mileage, 1957–74
('000m passenger-miles)

Year	Air	Rail	Road		
			Public service vehicles	Private transport	Total
1957	0.3 (0.2%)	25.9 (19.7%)	45.5 (34.6%)	59.9 (45.5%)	131.6
1965	1.0 (0.5%)	21.8 (10.6%)	39.2 (19.0%)	144.7 (70.0%)	206.7
1974	1.5 (0.5%)	22.4 (8.4%)	34.2 (12.4%)	217.5 (78.9%)	275.6

Source: After Prest and Coppock (eds), *The UK Economy*, p. 197.

that the majority of the population lived in rural areas. Manufacture was largely carried out in very small enterprises, often in cottage industries where manufacture was combined with agricultural activities. Much early production was related to agriculture: charcoal was needed for the iron industry, textiles were dependent on wool, brewing, bootmaking, saddlery, building, furniture making—all these trades use agricultural products as their primary inputs. Coal, iron and tin mining were the major exceptions, but even these activities were carried out in small village communities.

Foreign trade was orientated towards supplementing what British agriculture could not supply. Timber, hemp, tar and other vital supplies for the shipyards; tea, coffee, cocoa, tobacco and spices for newly refined palates; calico, silk and linen to supplement coarse woollen cloth.

Although British society in the middle of the eighteenth century may be described as rural, no Arcadian meaning must be imputed to the word. Much economic life was indeed concentrated in the many villages and hamlets of England and much manufacturing activity occurred in these villages. But life was by no means idyllic. The landless agricultural worker, the weaver, miner, blacksmith, spinner, tanner or whatever they were, probably led a harsh life dominated by poverty, hard labour, subordination, disease and early death. The village communities may well have had that community spirit which we now find so elusive, but that spirit often was more like that of a hungry citadel under siege than the cheerful association of carefree people that we now like to imagine.

Less than a hundred years later, by 1850, the social and economic structure of Britain had radically altered. Large wholly-mechanised industries had developed and many of these were no longer related to domestic agricultural production. Even the textile industry was now largely based on international trade in that it imported much of its raw material—more specifically cotton—and exported much of its product. The iron industry, which we might by then be justified in calling the iron and steel industry, had broken its dependence on charcoal and used coal instead. All industries had broken their total dependence upon natural streams and rivers. The source of power was steam and a network of canals, roads, and most recently railroads, had been established and was rapidly expanding. The centre of economic activity had decisively shifted towards the city. The majority of the vastly increased population had become city dwellers and urban proletarians.

By the middle of the nineteenth century Britain had achieved all the essential attributes of an industrial society and had become the first industrial nation. The transformation from a rural to an industrial society in the span of perhaps as few as seventy years (most writers consider 1760 to 1830 as the period of transition), is what we mean by industrial revolution. Other countries followed the industrial path only a few years later, with substantial differences in approach and conditions, but with essentially similar results.

By way of illustration, let us look at a few economic and demographic indicators to provide something of a grand overview of the differences between rural and industrial Britain (see Tables 1.6 to 1.9). The transformation of British society from agrarian to industrial—the industrial revolution—was not completed by 1830, but the impetus had been given and rapid further industrialisation and mechanisation proceeded almost inevitably. The industrial revolution has been described in great detail in a vast and often excellent literature and need not detain us for long.[4] What matters to us is the all embracing nature of the transformation—a complete social change by and through the use of new technology—and the multifarious nature of the forces which brought about the changes. For, far from

Table 1.6. United Kingdom pig iron output (tons)

1740	17,350	1830	678,417
1788	68,300	1835	940,000
1796	125,079	1839	1,248,781
1806	258,206	1848	1,998,568
1825	581,367	1852	2,701,000

Source: After Landes, D. S., *The Unbound Prometheus*, New York, Cambridge University Press, 1969, p. 96.

Table 1.7. Growth in railways and shipping

Decade	Average per annum registered in UK (thousand tons)		Year	Miles of railway track
	Sailing ships	Steam ships		
1790–99	1,443	—		
1800–09	2,003	—		
1810–19	2,378	1		
1820–29	2,291	17		
1830–39	2,278	52		
1840–49	3,110	122	1845	2,441
1850–59	3,867	319	1850	6,084
1860–69	4,690	724	1860	9,069
1870–79	4,240	1,847	1870	13,562
1880–89	3,440	3,783	1880	15,563
1890–99	2,784	5,993	1890	17,281
			1900	18,680

Source: After Mathias, P., *The First Industrial Nation*, London, Methuen, 1969, pp. 488–9.

seeking simple cause and effect relationships, we must look at a constellation of circumstances, a field of forces, which in their multiple interactions determined the course of history. Not seeking simple cause and effect relationships does not lead to obfuscation, but acknowledges the essential complexity of social developments and attempts to avoid gross falsification by the imposition of an alien *post hoc* logic. We shall seek to clarify the role of different factors and forces in their mutual dependence, rather than seek single causes for observed effects. As Peter Mathias observes:

To search for a single-cause explanation for the industrial revolution is to pose a false analogy with a simple equation governing chemical change. It is less tidy, less satisfying, less simple, but nevertheless more accurate to suppose that there was no one secret key which undid the lock, no single operative variable, no one prime relationship which had to be positive and in terms of which all other aspects of change may be regarded as dependent variables.[5]

The sequence and pattern of change is too complex to have anything as simple as a cause. We have to view the causation as a constellation of many circumstances and actions which between them favoured a certain progression of events. Eventually, the events added up to what in retrospect may be called the industrial revolution. Saying that only a constellation of circumstances can cause very complex systems to change is not to deny that there may be a few simple driving forces, such as acquisitiveness, greed, search for power and distinction, need for security, or whatever basic human attributes one wishes to invoke. However, human attributes change but little and slowly, while society during revolutionary periods changes very rapidly. The latent driving forces unleash actions which they were unable to achieve before and the conditions which make this happen are the explanations we seek. What were the triggers that unleashed the actions? What were the strands

Table 1.8. Average annual United Kingdom production of coal and imports of raw cotton

Decade	Raw cotton imported (thousand lb.)	Year or quinquennium	Coal produced (million tons)
1750–59	2,810		
1760–69	3,531		
1770–79	4,797		
1780–89	15,511		
1790–99	28,645		
1800–09	59,554	1800	11.0
1810–19	96,339	1816	15.9
1820–29	173,000	1820	17.4
1830–39	302,000	1850	49.4
1840–49	550,000	1855–59	66.7
1850–59	795,000	1860–64	84.9
		1865–69	103.0
		1870–74	121.3

Source: After Mathias, *The First Industrial Nation*, pp. 481 and 486.

Table 1.9. Growth and distribution of population in Great Britain

Total population (millions)

1690	(estimate 5.5, England and Wales)		
1750	(estimate 6.0, England and Wales)		
1801	10.69	1871	26.16
1811	12.15	1881	29.79
1821	14.21	1891	33.12
1831	16.37	1901	37.09
1841	18.55	1911	40.89
1851	20.88	1921	42.81
1861	23.19		

Population per square mile in six most densely populated counties

	1700	1750	1881
Middlesex	2,221	2,283	10,387
Surrey	207	276	1,919
Gloucester	123	157	
Northampton	121		
Somerset	119		
Worcester	119	148	
Warwick		159	825
Lancashire		156	1,813
Durham			891
Stafford			862

Population in selected cities (thousands)

	1760	1801	1841	1881	1921
Birmingham	28–30	71	202	546	919
Bristol		61	124	207	377
Glasgow		77	287	673	1,034
Liverpool	30–40	82	299	627	803
London		1,117	2,239	4,770	7,488
Manchester	30–45	75	252	502	730
Norwich	40–60	36	62	88	121
Sheffield	20–30	46	111	285	491

Sources: Mathias, *The First Industrial Nation*, pp. 449 and 451; Toynbee, A., *Lectures on the Industrial Revolution of the Eighteenth Century in England*, London, Longmans, 1906, pp. 32–6.

of circumstances which caused those momentous social, economic and technical changes known as the industrial revolution?

Without claiming any completeness of description, for such claims are always dangerous, the strands we must consider are:

  (i) increased efficiency of agricultural production;
  (ii) accumulation of capital through trade and agriculture;
  (iii) an accumulation of manufacturing experience and technical knowledge through craft skills;
  (iv) adequate supplies of energy and raw materials, especially coal and ores;
  (v) an adequately flexible political and social system;
  (vi) the great mechanical inventions, such as textile machinery, the steam engine and the railways;
  (vii) an aggressively adventurous spirit.

The period of the industrial revolution coincides with a period of rapid growth of population, presumably because a degree of wealth and of sanitation had been achieved which reduced death rates below birth rates. To feed the growing population without excessive demands on labour—thus freeing a labour surplus for industry—was an essential condition for industrialisation and required increased efficiency of agricultural production.

The new industries required large investment and this means that economic surpluses, beyond subsistence level, had not only to be created but also had to be channelled to industrial investment. Putting together capital—industrial investment—and labour, not to mention large supplies of dependable energy in the form of coal, creates the essence of industry; the only further ingredient needed to cement it all together is know-how. Unless a supply of skilled craftsmen is among the labour force, no combination of capital, energy and labour can achieve rapid technical progress.

Finally, the possibility of directing economic surpluses and labour in new directions presupposes social flexibility. Workers must be willing and able to move to where work is available and the wealthy must be willing to invest and risk money in new ventures. Social mobility and social rewards must be compatible with new requirements and must support the entrepreneurial and technical skills in their new enterprise.

Under these circumstances the great mechanical inventions could flourish and eventually led to the industrial growth which has continued to the present day. One of the hallmarks of the industrial age is the use of 'artificial forces, such as steam or electricity; these are tractable, regular and indefatigable, and can be increased indefinitely and at will.'[6]

CHANGES IN BRITISH AGRICULTURE DURING
THE INDUSTRIAL REVOLUTION

The reason most commonly given for the rapid improvement in the efficiency of agricultural production during the period of the industrial revolution is the increased rate of enclosure. Up to the beginning of the eighteenth century, most land in England was used on the common system, whether for pasture or cultivation. The method involved small strips of land within a large field being cultivated by different farmers. This meant a great deal of wasted effort in moving from strip to strip, and a disincentive to improve methods of tilling as the whole plot could only be as good as the poorest strip in it. Any adventurous new crops were ruled out as the agreement of all the cultivators had to be achieved before any change could be made. The land was usually owned by relatively large owners, but most of the work on this land was done by tenant farmers and their labourers, and in the case of small freeholdings by the farmers themselves, the yeomen.

The enclosure of land, meaning the concentration of all the strips cultivated by one farmer into unified fields, occurred from time to time by Act of Parliament. Between 1710 and 1760 the enclosures covered 334,974 acres. The pace of enclosures increased rapidly after 1760 and in the next eighty-three years nearly 7,000,000 acres were enclosed.[7]

The concentration of land ownership was very great. In mid-eighteenth century England half the land was in the hands of 2,512 persons.[8] Enclosures did nothing to reduce the concentration of ownership and indeed coincided with the almost total destruction of the yeomanry. Cultivation became much more efficient with enclosures and the small freeholder found it harder to compete against the large tenant farmer. But perhaps more important was the desire by the newly rich industrialists to join the politically and socially privileged land owners. There was a great demand for land and some of this demand was satisfied by the sale of yeomen's land.

The decline of the cottage industries dealt further blows to the yeomen, who were often involved in weaving and other small-scale industries. Not only could they not compete against factory-made goods, they also lost their ready marketing facilities by the decline of the old system of direct selling in the formerly flourishing market towns.

The fact that by the middle of the eighteenth century the majority of rural workers were wage earners and that the freehold farmers had largely lost their possession of the land was a factor of some importance. For only wage earners are free to shift into newly developing areas of economic activity; peasants are rooted to the land and any decline in agricultural prices simply pushes them deeper into poverty, without making them readily available for alternative work away from their land. The wage earner, unlike the peasant, gains from lower food prices and is more free to follow earning opportunities wherever they may arise. Even without a decline in the number of agricultural workers, the increase of population assured a steady supply of migrant labour for the new industries.

As the enclosures proceeded, so technical innovation in agriculture began to spread, albeit slowly at first. The landowners were often enterprising people and so were many of the tenant farmers. The larger plots of land made rationalisation of production easier and each farmer was now in full control of his own production methods. The developing industrial system and the rising spirit of enquiry provided further impetus to the introduction of new farm machinery, better methods of crop rotation, better drainage, new and improved varieties of crops and breeds of animals. Technical progress in agriculture was favoured by several factors in addition to enclosures. Marginal producers were mostly removed and the larger farmers could generally find sufficient capital for innovation, sometimes by the injection of capital into rural ventures by newly rich industrialists. The fact that surplus labour could migrate to the cities and be absorbed in the newly developing industrial and service sectors of the economy made the introduction of labour-saving methods in agriculture socially and politically possible. A further factor was the economic need to produce more and more food for a population that was not only growing rapidly but also increasingly consisted of city dwellers. Thus a decreasing proportion of the workforce had to produce sufficient food for an ever larger population. This was aided by improved transport and therefore marketing possibilities. As roads, canals, and eventually railroads improved, so it became possible to supply cities from distant producers and vice versa—large efficient producers were not restricted in their output by purely local demand. The final factor was the availability of inventions. The age produced the first fruits of scientific thinking and of engineering developments, which became as important to agriculture as to industry itself.

The first major innovation in agriculture consisted of new rotation patterns and new crops. Clover, lucernes and other new pasture crops helped to eliminate the need to leave land fallow from time to time as these crops 'fix' nitrogen and thereby enrich the soil. The possibility of bringing fallows into production helped not only to increase output when demand was high, but also provided flexibility to adjust output to market requirements.[9]

Flexibility and great reductions in the need for permanent pasture were also achieved by a range of new fodder crops, in particular turnips and swedes. Smaller areas of land were thus required to feed a larger number of animals and thereby more food crops could be grown without loss of fertility, which was then almost exclusively provided by animal manure.

Many of the first innovative ideas in agriculture originated in the Netherlands and were brought to England by travellers and exiles. The principle of cross-fertilisation of ideas worked here as elsewhere. Imported and indigenous ideas mixed readily and farming methods improved, slowly at first but at an accelerating pace towards the middle of the nineteenth century. Contrary to popular belief, agricutural employment did not reach its peak till about 1850.

Not only crops and rotation patterns, but also farm implements changed substantially. One of the better known innovators in this field was Jethro Tull, whom Peter Mathias describes as

being typical of the experimental and scientific tradition of his day: a mixture of brilliant observation and false hypotheses . . . He measured and controlled his experiments, establishing a tradition of experiment and observation in implements as well as in technique. . . . In particular, he advocated the use of efficient light iron implements—harrows, rakes and hoes—drawn by a single horse, publicising this in his famous book, 'Horse-hoeing Husbandry' (1733).[10]

The horse generally replaced the slower ox, and new types of plough, seed drills and, eventually, threshing machines set the pattern for the new agriculture. The pacemakers were the counties of eastern and central England, with their flat fields of light soil and relatively long growing season.

Improved crops and methods of cultivation were supplemented by improved breeds of livestock and better animal husbandry. Better feedstuffs paved the way for these improvements, but no doubt the general spirit of improvement and experimentation, aided considerably by greater ease of travel, as well as the new large markets, were important factors. Of primary economic importance were new breeds of sheep, such as the Leicester, and new breeds of cattle, such as the Durham Shorthorn. But although the primary economic function of the racehorse has always been in the leisure industry, it also served as an important testing ground for animal breeding.[11]

Despite the increases in agricultural efficiency and agricultural production, Britain became a net importer of wheat in the fourth quarter of the eighteenth century. The reasons must be sought in a greatly increased population and presumably in increased meat consumption, which decreases the overall food-growing capacity of a given area of land.[12] The modern pattern of British trade—the import of food and raw materials in exchange for manufactured goods and services such as shipping and insurance—thus started to develop in the early stages of the industrial revolution.

EARLY MANUFACTURE AND THE ACCUMULATION OF CAPITAL

England before the industrial revolution certainly was not entirely devoid of manufactured goods. The main products were woollen cloth and iron. Woollens reached an export value of £4 million in 1770 and constituted nearly a third of total exports.[13] In 1740 there were ten iron furnaces in Sussex, producing 1,400 tons while Gloucestershire, Shropshire and Yorkshire had six furnaces each. In 1737 there was a total of fifty-nine furnaces in eighteen counties, producing 17,350 tons of iron annually. An approximately equal, or slightly greater, amount of iron was imported. Not quite 150 years later, in 1881, exports of iron and steel exceeded 3.8 million tons and were valued at £27.6 million. The value of imports, £3.7 million, was by then a mere fraction of exports.[14]

The manufacture of hardware, chiefly for domestic use, was concentrated in Birmingham and Sheffield, with quite a few smaller workshops in many parts of the country. In 1727 there were about 50,000 people engaged in the hardware trade in Birmingham alone.

In 1760 the cotton industry was a cottage industry of tiny proportions. The total number of people engaged in cotton manufacture was estimated as 40,000 and the annual value of their product was a mere £60,000. The only important mechanical innovation of the pre-industrial period in textiles was the fly shuttle invented by John Kay in 1738, which doubled the productivity of weaving looms.[15] Other manufactures of some significance were silk hosiery, woollen hosiery, ribbons and linen. 'The "manufacturer" was, literally, the man who worked with his own hands in his own cottage.'[16] The combination of agriculture and manufacture was common and many small freeholders supplemented their meagre income by weaving or other manufacture. Sometimes these small manufacturers would employ quite a few workers—men, women and children.

Roads were appalling and waterways were only slowly developing. Some river-beds were widened and deepened between 1660 and 1755, when the first eleven miles of canal were built near Liverpool.

The building of canals spread rapidly after 1760. Trade in the first half of the eighteenth century was largely restricted to local marketing, though there was some trade between counties. The selling of goods was a laborious affair of trekking round markets and fairs by travelling salesmen—chapmen—and their pack horses. Nevertheless, by the middle of the eighteenth century many merchants had accumulated some capital and were employing it to start a process we would now call 'vertical integration'. Instead of buying other people's products and selling these, they would employ cottagers to manufacture for them. The merchant became a 'putter out', supplying raw materials to the weavers, collecting cloth or hosiery from them, and marketing these goods in market towns. The small merchant was using his accumulated capital to become a forerunner of industrial capitalism. 'Finally, the merchant would get together 30 or 40 looms in a town. This was the nearest approach to the capitalist system before the great mechanical inventions.'[17]

In addition to the small merchants there were, of course, large traders involved in overseas trade. The size of ships was growing and by 1770 ships of 900 tons were not uncommon. The total value of exports in 1760 was £14.5 million, with the colonies taking about one third of these.[18] Trade was strongly supported by government mercantilistic policies, attempting to maximise export earnings. The large trading companies, such as the East India Company, enjoyed considerable power and protection. Some of the capital of the large and small traders eventually became available for investment in manufacture and new transport services.

Savings on a wider scale, at first mainly from agricultural sources, became available for industrial investment with the development and growth of the banking system. By the time the industrial revolution was getting under way, the banks had developed sufficiently to enable them to channel savings from agriculture into investment in transport and industry. Banks proved particularly vital for the provision of short-term working capital for industry, which is every bit as important as long-term investment into fixed assets. Demand for capital increased continuously, quite slowly at first, and became well-nigh insatiable with the development of railways at the very end of the industrial revolution proper. In fact we must see the industrial revolution as the beginning of industrialisation, the passing of a threshold, and certainly not as its completion. So, for example, the Stock Exchange, long established for the financing of government borrowing requirements, only began to play a significant role in raising industrial capital from about the middle of the nineteenth century.

A major source of finance was, and is of course, the profit ploughed back into the business. Many early manufacturers put the growth of their enterprise before personal consumption and used much of the available income for further investment. Often not only the industrialist but also his friends and family proved important sources of finance: dowries and various settlements; mortgages on land and property; all these were of significance in the early, highly personal, capitalist system. Quite frequently the industrial entrepreneur had first accumulated capital in a related branch of business.

Many grain merchants or maltsters became brewers; the characteristic industrialist in the woollen industry had set up a factory from being a woollen merchant. . . . In the metal industries most of the iron masters had been in the secondary metal trades—making final products from refined metals—and they then moved back to making iron. . . . Apart from capital, such links gave knowledge of the trade, help with orders and friends in useful places.[19]

THE GREAT MECHANICAL INVENTIONS

One of the major strands in the story of industrialisation undoubtedly concerns the development of the machines which formed the very heart of the industrial system. Although inventions themselves cannot suffice to bring about industrialisation, they do form a vital aspect of it. While we must avoid the trap of thinking that the mechanical inventions brought about the industrial revolution in the manner that the gravitational pull of the moon brings about the tides, inexorably and necessarily, we must not go to the other extreme of thinking of the industrial revolution merely as a political and economic development. The industrial revolution is inextricably linked with the great inventions of the period, even if the links are more complex and varied than simple cause and effect relationships.

The three main groups of inventions that determined the shape of the first industrial revolution were: textile machinery; steam engines and steam powered transport; and improvements in iron and steel manufacture.

The origins of the steam engine go back considerably before the industrial revolution and the early names associated with it are Papin in France and Thomas Savery in England. Papin's name is remembered on the continent of Europe mainly by its association with the pressure cooker, while Savery seems to have only a few admirers left and these are concentrated in British academic circles.

Savery's own claims to fame were somewhat uninhibited but form fascinating reading. Savery did not produce a 'prime mover', but a steam pump for which he claimed far greater efficiency and capacity than the horse-driven pumps used in the mines of the early 1700s could achieve. The pump essentially consisted of a steam boiler which would fill a vessel with steam. The vessel would then be cooled, thereby condensing the steam and creating a vacuum which would draw water to a height of up to 80 feet. Savery was well aware of the economics of coal mines and of the value of entrepreneurial incentives. 'The cheaper Water is drawn, the more is the Miner incouraged to adventure . . .' A social dilemma which Savery soon encountered was fierce opposition to his pump by the carpenters who made the traditional horse-driven pumps. The steam pump required some metal work and not nearly as much carpentry, hence an early version of the argument about losses of jobs and skills arose.[20]

The first steam device that might properly be called a steam engine was designed by John Newcomen between 1705 and 1711. The pumping action was not by vacuum as in Savery's device, but by a steam-driven piston. A steam boiler was used to fill a cylinder with steam. The cylinder was then cooled and the steam thereby condensed. Thus the pressure in the cylinder became lower than atmospheric and the atmospheric pressure exerted on the piston therefore caused it to move into the cylinder. The piston was connected to a beam and the beam to another piston, acting as a water pump. When the cylinder was refilled with steam, the piston moved out again and the cycle of operations was complete. The engine was known as an 'atmospheric engine', as the pressure of the atmosphere was the moving force.

Apart from its extensive use as a pump in mines, the Newcomen engine also achieved some use for pumping water to drive water-wheels. Many of the early mechanical looms and spinning machines, not to mention forges and their mechanical hammers, were driven by water-wheels and hence situated near streams. To gain some independence from the vagaries of water levels in the streams without a radical change in technology, Newcomen engines were sometimes installed to drive the water-wheels. Similarly, one of the early uses of steam engines was to drive bellows for blast furnaces, thus demonstrating the many interdependencies of the industrial system: steam engines needed iron vessels, cylinders, pistons and other parts; blast furnaces needed steam engines to produce iron. . .

The most famous name of all, often mistaken for the sole inventor of the steam engine, is James Watt. Watt was an instrument mechanic at Glasgow University and serves almost as a symbol linking the craft traditions of pre-industrial manufacture with the scientific foundations of nineteenth and twentieth-century engineering. Watt was a craftsman, yet Watt was also the friend, collaborator and equal of the scientific establishment in the University.

Watt was given the task of repairing a Newcomen engine and was struck by the inefficiency of constant heating and cooling of the cylinder. No doubt he was aware of the nature of heat capacity and realised that much greater efficiency could be achieved if the cylinder were permanently hot and the condenser permanently cold. The obvious solution was to separate the two functions into different vessels, and this is the essence of Watt's first great contribution to engineering science.

After several years of experimentation Watt obtained his first patent for a steam engine with a separate condenser in 1769. Both for his experiments, but even more for his attempts to exploit the invention and produce steam engines, Watt needed money. He found an early backer in Roebuck, but eventually Roebuck went bankrupt and became unable to back Watt any further. Through the mediation of some of his friends in Glasgow, Watt was introduced to the wealthy and entrepreneurial Birmingham industrialist Matthew Boulton.

Neither Boulton nor Watt was fully aware of the extent of the difficulties and financial commitments ahead. Had they known, it is doubtful whether the venture would ever have started. Perhaps the eventual achievements surpassed all their wildest dreams, but undoubtedly the difficulties surpassed all their most frightful nightmares.

Boulton's Soho Works in Birmingham, a general metal working factory, set about to manufacture, sell, use, and improve Watt's steam engine. The first engines were used as straight replacements for Newcomen's pumps. Only in 1775 came the first order, from the famous iron master John Wilkinson, for a steam engine to drive machinery directly. Many further steps, inventions and developments were necessary before the steam engine could become the universal prime mover of the first industrial age.[21] One of the early problems was the accurate boring of the cylinders, so that the piston would fit with sufficient precision to avoid, at least when packed with rope, escape of steam. John Wilkinson had worked out a method of boring cannon and this technique was transferred to the manufacture of steam engines.[22] Another major contribution by Watt was the development of a system to convert the to-and-fro motion of the piston into the rotary motion required to drive much factory machinery. Watt used planetary gears and obtained a patent in 1781.[23]

With hindsight, the diffusion of steam engines seems inevitable and the dominance of Watt engines over their predecessors appears pre-ordained. But whatever the social savings[24] of such engines may be, individual decisions on their purchase and introduction had to be made by individual factory owners. Their decisions were based on an assessment of the standing of their own firm in the world and an appraisal of the problems they had to solve in the light of solutions on offer. If a factory was expanding rapidly and the water power from its streams became inadequate, then clearly a steam engine was required. Which engine to purchase would depend on initial capital outlay, estimated running costs, and the general reputation of the different engines. In fact the diffusion of steam engines proceeded quite slowly and even Savery and Newcomen engines were not displaced over night. The hundreds of individual decisions only gradually shifted the pattern of diffusion of steam power and as it shifted, further decisions became easier because Watt's engine had developed and gained many advantages and the growth of industry became firmly established. With *post hoc* rationalisation it is possible to discern Adam Smith's invisible hand, for example when it appears to have directed large consumers of power, such as the cotton industry, to areas of cheap coal, as shown in Table 1.10. Or could it have been the other way round, that cheap coal supplies followed the industry?

First the greatly improved efficiency of Watt's engines compared to Newcomen's, then the rotary engine combined with good manufacturing facilities and erection services, provided by the Boulton-Watt partnership, paved the way towards the provision of a reliable, economic, convenient and inexhaustible source of energy, at least

Table 1.10. Location of textile mills, 1838

	Total steam horsepower	Coal less than 10s. per ton %	Coal between 10 and 20s. per ton %	Coal over 20s. per ton %
Cotton	40,783	96	4	<0.5
Worsted	5,863	86	12	2
Woollen	10,887	81	19	1
Flax	3,134	69	29	2
Silk	2,320	54	33	12

Source: After Tunzelmann, G. N. von, *Steam Power and British Industrialization*, Oxford, Oxford University Press, 1978, p. 66.

while coal supplies were available at reasonable cost. To provide similar power vast numbers of horses would have been required and the logistic and economic problems of using and feeding huge teams of animals would have been insuperable: 'in 1870 the capacity of Great Britain's steam-engines was about 4 million horsepower, equivalent to the power that could be generated by 6 million horses or 40 million men . . . On balance, double the number of men or animals would seem a more accurate estimate.'[25] But large as this figure is, let it not be assumed that steam engines spread like wild-fire. In fact even in 1850 more than half the power used in the woollen industry came from water, as did about one-tenth of the power in the pioneering cotton industry. Table 1.11 shows the spread of steam horse power and the simultaneous decline of water power in the cotton industry.

Table 1.11. Source of power in the cotton industry 1835–56
        for England and Wales

Year	Steam horsepower	Water horsepower
1835	36,000	
1838	50,300	
1842	52,400	9,677
1850	71,100	8,182
1856	86,400	6,551

Source: After Tunzelmann, Steam Power and British Industrialization to 1860, pp. 139 and 235.

In the United Kingdom as a whole in 1850 there was about 0.5 million horsepower in stationary steam engines and nearly 0.8 million in mobile engines, mostly railway locomotives.[26] The comparative advantages of steam over water, wind or horses became gradually overwhelming. As Landes points out, 'Coal and steam . . . did not make the Industrial Revolution, but they permitted its extraordinary development and diffusion'.[27]

The locomotive seems, with hindsight, an almost inevitable consequence of the steam engine. But it took many years of development of steam engines before they could be made small enough to even contemplate putting them on wheels. To make such engines possible, better iron products and better metal working methods had to be developed. No doubt the steam engine itself contributed to these developments.

Like all major innovations, the railway had its precursors. Wagons running first on wooden and later on iron rails had long been used to haul coal from pit-head to barge. In the age of experimentation with steam power, when horses were generally being replaced as a source of stationary power, the thought of replacing them in this mobile application was obvious. In some ways, Watt himself proved an obstacle to railway development because he obstinately refused to contemplate a high pressure engine, which could operate with smaller cylinders. Perhaps he was worried about the possibility of explosions, and if so his fears were often borne out by future events. Watt retained a monopoly position, by special Act of Parliament, till 1800.[28] After that, the field became wide open and a large number of new entrants rapidly advanced the technology. Similarly, a large number of people attempted to build locomotives, but only the winners of the famous Rainhill trial, arranged by the Liverpool and Manchester Railway Co. in 1829, George and Robert Stephenson reached fame and fortune. From the opening of the Manchester–Liverpool line the following year, progress of the steam-hauled railway seemed unstoppable. Despite somewhat lower freight rates on the canals the railway proved a formidable competitor for long distance haulage of bulk goods and even more so for consignments of high value goods. The speed, reliability and frequency of trains gave them a tremendous competitive advantage. The added bonus of passenger travel opened an entirely new era. The mobility of people increased vastly and this new found mobility made an enormous contribution to the spread of new ideas and the rapid build-up of a large industrial network.

Britain became railway mad. Railway companies mushroomed and absorbed large quantities of investment capital; £250 million in the twenty years from 1830 to 1850.[29] The impetus the railways gave the rest of the economy was extraordinary. Not only did they increase mobility, but their building provided work for civil engineers supported by armies of labourers; stations and depots had to be built; local haulage to the railway stations developed, locomotives and rolling stock were manufactured, signalling systems were developed. Vast amounts of steel went into rails and bridges, vast quantities of timber into sleepers. So great was the fillip to the economy that it has become fashionable to think of the technological innovation of the railways as the cause and point of departure of a long-term upswing in the economy.[30]

The development of textile machinery, the real backbone of the first factory system, is well known and only a few salient points will be picked up here. The leadership was definitely taken by the spinning operation, mainly in cotton, followed by wool. The main inventions were James Hargreaves' 'spinning jenny' of 1765–6, Richard Arkwright's 'water frame' of 1766–9, and Samuel Crompton's 'mule' of 1774–9. The water frame and particularly the mule, which underwent much further development, were large power-driven machines unsuited for a cottage industry. For example, Jedediah Strutt's mill in Belper required an investment of around £15,000 in 1793, of which almost one-third went into machinery, one-third into buildings and one-third into initial stocks of raw materials.[31]

The new mechanical spinning machines developed earlier than power looms, perhaps because the technical problems were relatively easier to solve, but perhaps also because weavers were crying out for more yarn. In the 1760s spinning represented the bottleneck in the industry, which was remedied only by about 1790, partly because of the introduction of the spinning jenny and the water frame.[32]

The introduction of large machines was one of the factors which caused the spinning industry to become the first industry to adopt the factory system of production. The merchants and putters-out had already developed a capitalist system of production, but the work process was still carried out in very small units. No doubt the desire for better control over the workers, the work-flow and the flow of goods had commended the advantages of centralisation to the putters-out. When these considerations came into confluence with the newly available large efficient machines, the factory system resulted. Organisational efficiency called for the factory system, while requirements of technical efficiency and productivity demanded highly mechanised spinning devices. When the latter proved to demand high investment and large-scale production, the former became irresistible. The factory system was thus born out of many circumstances, of which technological invention was but one, albeit important, item. The results were, of course, far beyond the reach or comprehension of the individual decision makers.

The example of the successful mechanisation of spinning further intensified the efforts towards technical improvements in weaving. The first truly successful power-loom was invented by Edmund Cartwright in 1785. The loom was converted from horse power to steam power in 1789 and, although weavers burned down Cartwright's factory in 1791, the invention was the precursor of the mechanised power-driven factory loom. The speed of introduction must not be exaggerated. What seems with hindsight an overnight inevitable change required a great deal of effort, money, struggle and time. In fact as late as 1834 there were twice as many hand-looms as power-looms and full mechanisation was not completed till about 1850.[33]

Tunzelmann has attempted to calculate the economic incentive for conversion from hand-looms to power-looms (see Table 1.12). The net benefits become large only after a number of years and thus the purely economic case is not immediately overwhelming. This was particularly so as 'the hand-loomer was prepared to reconcile himself to lower and lower wages rather than abandon his skilled trade'.[34]

Table 1.12. Costs and benefits of power-looms versus hand-looms for 50-reed cambrics in Bolton, mid-1830s (£)

	Year 1	Year 2	Year 13
Capital costs:			
(i) Power-loom	−9		
(ii) Dressing machine	−1.67		
(iii) Buildings	−7.27		+3.39
(iv) Power	−7.05	−1.01	+0.79
Material costs:			
(v) Flour	−1.16	−1.16	−1.16
(vi) Oil, lighting	−1.00	−1.00	−1.00
Labour costs:			
(vii) Dressing	−2.58	−2.58	−2.58
(viii) Overlooking	−1.31	−1.31	−1.31
(ix) Weaving	−5.44	−5.44	−5.44
(x) Extra output:	+14.57	+14.57	+14.57
(xi) Net benefits:	−21.91	+2.07	+5.68

Source: After Tunzelmann, *Steam Power*, p. 198.

The rise of the cotton industry had many far-reaching consequences. Not only did it establish the factory system of production, meaning large numbers of workers employed under the same roof and working under rigid discipline imposed by the owner and demanded by the power-driven machinery, but it also produced goods relatively cheaply and thereby created massive demand for them. Because cotton had to be imported, the cotton industry also became a key factor in a massive expansion of British foreign trade. The creation of an urban proletariat on the one hand and a new middle class of capitalist manufacturers, whose wealth was not based on the land, had far-reaching social and political consequences.

It has been argued that only because the new class was able to penetrate into the higher echelons of British society, and eventually rival the power of the landed gentry, did the industrial revolution proceed apace. Had the political and social system been more rigid, then perhaps the new source of wealth would not have been able to compete for power against the old source—land.

One of the many remarkable features of early industrialisation was that despite all the labour-saving effects of the new machinery, output grew fast enough not only to maintain, but in fact to increase employment. Then as now employment was a result of a balance between growth in productivity and growth in demand. As Table 1.13 shows, demand outstripped growth in productivity by a considerable margin in the early days of the mechanised cotton industry.

Table 1.13. Employment and output in cotton mills, 1835–56

Year	Employment			Quantity of cloth woven (m. lb.)	Quantity of yarn spun (m. lb.)
	Lancashire ('000)	England & Wales ('000)	United Kingdom ('000)		
1835	122	183	219	200	277
1838	150	219	259		
1842	164	232	271		
1850	216	293	331	421	571
1856	259	341	378	588	753

Source: After Tunzelmann, *Steam Power*, pp. 210, 239.

One further consequence needs to be stressed here. It became apparent very quickly that the industrial system is indeed a system. A factory in isolation cannot exist, for it depends on a large range of inputs of all kinds. The cotton industry needed to be supplied not only with raw materials, which needed to be transported, but also with buildings and with specialised machinery. The production of this machinery gave rise to workshops surrounding the cotton industry and these workshops themselves became engineering factories in due course. And again the engineering factories needed inputs of components and of higher and higher quality iron and steel. Though the first machines were still largely built of timber, such machines could not stand up to the pounding resulting from being driven by steam engines. Thus a demand for iron arose and also a demand for iron working skills, for iron working machinery and, eventually, for standardised components.

Cloth is not made by mechanical devices alone. Apart from spinning and weaving it requires bleaching and dyeing. And as the quantity and flow of production increased, so the deficiencies of the old methods of bleaching and dyeing became apparent and became the weak link in the production chain. In this way the cotton industry created a demand for new chemical methods and another constellation of circumstances, a confluence of demand from the cotton industry and new knowledge in chemistry became important stimuli to rapid growth in the chemical industry.[35] An interesting early example of a learning curve—the reduction in the cost of production of a commodity with accumulated experience—is the manufacture of bleaching powder. Demand creates experience, experience reduces costs, lower price creates demand.

Table 1.14. Production and price of bleaching powder at St. Rollox works

Year	Tons	Price per ton
1799–1800	52	£140
1801	96	130
1802	72	112
1805	147	112
1810	239	93
1820	383	60
1830	1,447	25
1840	2,383	26
1850	5,719	14
1860	7,459	11
1870	9,251	8.5

Source: After Campbell, W. A., The Chemical Industry, London, Longman, 1971, p. 65.

The final strand of technical developments which were at the heart of the industrial revolution is the production of iron and steel. Since time immemorial (probably eighteenth century BC), iron had been produced by mixing iron ore with charcoal in a furnace, burning the charcoal by allowing or blowing air into the furnace and removing the iron which resulted. This rather impure and brittle lump had to be improved by reheating in open charcoal fires and repeated hammering. From roughly the sixteenth century, the furnaces became larger and used bellows driven by water wheels. By using charcoal from hardwoods, the temperature in the furnace could be raised sufficiently to melt the iron. In this process it acquired a high carbon content and was quite brittle, but very suitable for casting into ingots and, in this form, became known as pig iron. The pig iron could be used either directly as cast iron or could be reheated in an open furnace to reduce the carbon content and then forged by water-driven hammers into malleable steel.[36] As demand for iron and steel increased, supplies of charcoal became a bottleneck in the production chain. Especially in Britain, with limited native production of charcoal, there was a danger of being squeezed out by Swedish competition. Other problems arose because of unreliable water supplies to drive bellows and forge hammers. Sometimes Newcomen engines were used to raise water for this purpose.

There were many attempts to smelt iron with coal, or rather coke, and success finally came to Abraham Darby and his son, Abraham II, in 1709, though the process took a long time to perfect and did not diffuse widely till after 1760.[37] The next important step came with inventions which allowed the pig iron produced by the blast furnaces to be refined into malleable steel, again using coke, and to be shaped into long bars of regular cross-section. The two inventions were made by Henry Cort in 1784. In the puddling furnace the pig iron was heated by a coke fire, but was not in direct contact with the coke, and was stirred (puddled) by iron rods so as to cause impurities to float to the surface, from which they were removed with ladles. The purified iron was considerably more malleable than pig iron. One of the ways of using this malleability was the rolling mill, the second invention of Cort. Iron and steel, one of the principal materials of the industrial revolution, could now be produced by its main source of energy, coal, and by its prime mover, the steam engine. At this stage it became true to say that manufacture consisted of using the earth's resources of fossil fuel to transform raw materials, particularly minerals and agricultural produce, into desired products. The means of achieving this was by the use of a mix of machines and human labour. Thus the main factors of production, buildings, machinery, raw materials, energy and labour, became established. The capitalist provided all of these except labour and we therefore often speak of just two factors of production, capital and labour. Probably an unfortunate custom.

Coke-fired blast furnaces with steam-driven bellows, puddling furnaces, and steam-driven rolling mills became standard means of manufacture of iron and the industry became more and more located near coal fields, which provided the bulkiest and principal ingredient. Even today the British steel industry is still largely located in the areas in which it became established about two hundred years ago.

The puddling process became obsolete with the invention of the various non-mechanical converters of pig iron into steel, particularly by Henry Bessemer in 1855-6. The so-called Bessemer converter was widely used, but could not deal with iron containing sulphur and/or phosphorus. A modification by Sidney Thomas and Percy Gilchrist in 1878 took care of this problem by suitable cladding of the converter. Chemistry had very much entered the art of the metallurgist by then. Another modification, which proved highly efficient, was introduced by a combination of the efforts of Friedrich and Wilhelm Siemens between 1856 and 1861, and Emile and Pierre Martin in 1864, to create the Siemens–Martin converter, first used by Krupp in Essen in 1869. Figure 1.1 shows how the different methods of steel production rose and declined, while total world production increased continually and spectacularly. The rise and fall of techniques shown in Figure 1.1 is one of the most characteristic features of technology.

INDUSTRIALISATION OUTSIDE BRITAIN

If imitation be a form of flattery, then the British technical innovators had much cause to feel flattered. For the story of the industrial revolution in Britain is, with delays and variations, the story of early industrialisation in Belgium, France, the German states, Austria–Hungary and North America. Belgium was probably fastest in emulating British experience, probably aided by an abundance of coal and the proximity of British expertise. The related industries of mining, steel and engineering developed rapidly in the nineteenth century. France was somewhat slower, perhaps because of a relative shortage of coal, but probably also for political reasons. Germany developed more slowly still. The obstacles there consisted of a feudal structure and very powerful influence of the guilds. The combination of a large number of small states, powerful landowners, serfdom and guilds proved a strong disincentive to industrialisation, which only proceeded apace after the events of 1848 and subsequent years, which broke up the rigid political and social structure. When eventually the German states united under Prussian leadership in 1871, industrialisation truly forged ahead. Table 1.15 shows the development of both coal and steel production in some of the leading industrial countries in the second half of the nineteenth century.

Early industrialisation provides some fascinating examples of attempts by the state to influence and accelerate industrial

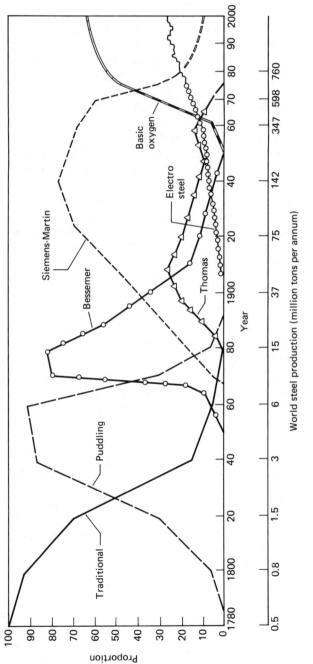

Fig. 1.1. Change in steel production methods. *Source:* After Roesch, K., *3500 Jahre Stahl*, München, VDI–Verlag, 1979, p. 48.

Table 1.15. Annual coal and steel production (million tons)

	UK		France		Germany		USA	
	Steel	Coal	Steel	Coal	Steel	Coal	Steel	Coal
1860–4		84.9		9.8		15.4		16.7
1865–9		103.0		12.4		23.5		26.7
1870–4	0.5	120.7	0.1	15.1	0.3	31.8	0.1	43.1
1875–9	0.9	133.3	0.3	16.3	0.4	38.4	0.6	52.2
1880–4	1.8	156.4	0.4	19.3	0.8	51.3	1.6	88.7
1885–9	3.0	165.2	0.5	20.7	1.1	60.9	2.8	115.3
1890–4	3.2	180.3	0.8	25.4	2.8	72.0	4.3	153.3
1895–9	4.2	201.9	1.3	29.6	5.1	89.3	7.6	189.1
1900–4	4.9	226.8	1.7	31.8	7.3	110.7	13.4	281.0

Source: After Ashworth, *A Short History of the International Economy Since 1850*, pp. 21–2.

development. While in Britain the role of the state was mostly confined to strong mercantilistic measures and commercial as well as political and military imperialism, not to mention the industrial incentives provided by the Napoleonic wars,[38] other states attempted some degree of planning and deliberate infrastructural support. Austria–Hungary may serve as an example.

Aristocratic and middle-class *Fabrikanten* were aided by loans, tax exemptions, and the like, and if necessary excused military service and guaranteed freedom of worship. Many of them had been induced by the Court to go into such ventures, particularly a number of Bohemian lords; others had been enticed from Germany, the Netherlands, or Switzerland. Several of the larger firms were taken over by the government when they ran into too many deficits, but this practice was discontinued towards the end of the mercantilist period. For increasing the supply of industrial labour, at the various levels of skill and technical education, the Austrian government went to great lengths to attract craftsmen and technicians from abroad (including England, Italy, and Silesia), promising them not only high wages but also housing, exemption from conscription, and similar benefits. Furthermore, by means of instructors and special 'schools' improved techniques of spinning, weaving, and knitting or lacing were diffused in the rural areas. Most important, perhaps, were the exemptions granted to the new industrial establishments from the rules and supervision of the guilds.[39]

The measures described above formed part of a policy of achieving economic autarky within the monarchy and were supplemented by strong protectionism. Other countries, with considerable local

differences, pursued essentially similar policies of support for industrialisation. Included in this support were measures of subsidies and
incentives for investment in railways, which in most countries were
subject to central planning of a national network. This was regarded
as necessary both to foster internal and external trade and to secure
mobility for the armies and their logistic support.

The total mileage of the world's railways increased from 4,772
miles in 1840 to 490,974 miles in 1900 and 795,213 miles in 1930.
The British system grew purely by private enterprise and comprised
6,621 somewhat haphazard miles in 1850. Learning from British
experience, Belgium planned a basic rail network in 1834, which
was completed in ten years and later grew by private accretion. In
France a mixture of private enterprise and state planning and assistance grew up from about 1837 and through the 1850s. A similar
mix of state intervention and private enterprise, though with many
changes of policy, occurred in Austria–Hungary, while in the German
states the railway system supplemented the customs union as an
instrument of unification. A milestone of transcontinental travel
was reached with the inauguration of the famous Orient Express
from Paris to Constantinople in 1888.[40]

The development in the United States provides a particularly
interesting illustration of a mix of economic and political motivations which played such major roles in railway construction.

The unique phase of American railway building came after the Civil War. In the
eastern part of the country, as in Europe, railways had been built to meet an
immediate demand, but the penetration by the railways of the lands between
the Mississippi and the Rockies occurred while the region was still economically
undeveloped and very sparsely inhabited. There could be no strong effective
demand for transport until the land was settled, but settlement on a large scale
was impossible until the railways provided easy access. The Union Pacific and
the Central Pacific, which together provided the first trans-continental route,
received their charters in 1862, and, with the aid of lavish land grants and loans
from the federal government, were able to complete construction in 1869.[41]

By the time of the first world war, all Western and Central European countries, to some extent Russia, and certainly the United
States, had become industrial nations. Urbanisation had proceeded
apace, though perhaps not to the same extent as in Britain; steel was
manufactured on a large scale; there were extensive networks of
railways and fully mechanised manufacture of machinery and consumer goods of all kinds. The steam engine had by then been supplemented by the internal combustion engine and the electric motor,
and the telephone and telegraph had supplemented communications.

Food processing, textiles, glass, paper had all become industrialised and the chemical industry had grown to major proportions. The millions of men slaughtered in the First World War bore dreadful witness to the industrial might of the warring nations.

Table 1.16 and Figure 1.2 sum up world industrial production from 1850 to 1979. The trend is one of considerable growth, but superimposed upon this trend are very large fluctuations of irregular periodicity and amplitude, with the depression of the late twenties and early thirties particularly marked.

Table 1.16. World industrial production ($100 million in 1913 prices) and energy consumption (in million ton coal equivalent)

Year	Industrial production	Energy consumption
1740	2.1	
1750	2.7	
1760	2.5	
1770	3.1	
1780	3.2	
1790	5.4	
1800	7.5	
1810	8.9	
1820	11	
1830	17	
1840	24	
1850	31	
1860	53	134.2
1870	68	208.5
1880	94	326.8
1890	144	510.1
1900	205	767.1
1910	302	1,200.4
1920	326	1,455.2
1930	448	1,693.1
1940	651	2,138.8
1950	812	2,647.8
1960	1,358	4,415.7
1970	2,335	7,150.2

Source: After Haustein, H. D. and Neuwirth, E., Long Waves in World Industrial Production, Energy Consumption, Innovations, Inventions, and Patents and their Identification by Spectral Analysis, Laxenburg, International Institute of Applied Systems Analysis, Working Paper 82-9, January 1982, Appendix A.

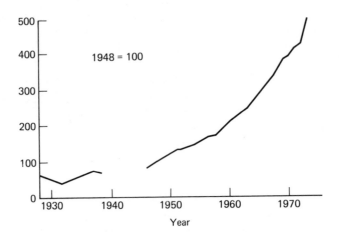

Fig. 1.2. World industrial production. *Source*: After Braun E. and Collingridge, D., *Technology and Survival*, London, Butterworths, SISCON series, 1977, p. 1.

NOTES

1. For a fuller discussion see Braun, E., 'Homo Faber, Homo Ludens and the Future of Work', pp. 353–61, in H. Bucholz and W. Gmelin (eds), *Science and Technology and the Future*, Munich, K. G. Saur, 1979 and Huizinga, J., *Homo Ludens*, London, Temple Smith, 1970.

2. See, for example, Marx, K., *Das Kapital*, English edition, London, Lawrence and Wishart, 1954.

3. See, for example, Bundesminister für Forschung und Technologie, *Politik Wertewandel Technologie*, Düsseldorf, Econ, 1982.

4. From the extensive literature on the industrial revolution and subsequent developments the following works have provided most of the material, though not the interpretations, here used: Ashworth, W., *A Short History of the International Economy since 1850*, London, Longmans, second edition, 1962; Best G., *Mid-Victorian Britain 1851–1875*, London, Weidenfeld and Nicolson, 1971; Checkland, S. G., *The Rise of Industrial Society in England 1815–1885*, London, Longman, 1964; Hartwell, R. M. (ed.), *The Causes of the Industrial Revolution in England*, London, Methuen, 1967; Hobsbawm, E. J., *Industry and Empire*, London, Weidenfeld and Nicolson, 1968; Kranzberg, M. and Pursell, C. W. (eds), *Technology in Western Civilisation*, New York, Oxford University Press, 1967; Landes, D. S., *The Unbound Prometheus*, New York, Cambridge University Press, 1969; Lane, P., *The Industrial Revolution*, London, Weidenfeld and

Nicolson, 1978; Mantoux, P., *The Industrial Revolution in the Eighteenth Century*, London, Jonathan Cape, revised edition, 1961; Mathias, P., *The First Industrial Nation*, London, Methuen, 1969; Russell, C. A. and Goodman, D. C. (eds), *Science and the Rise of Technology since 1800*, Bristol, John Wright, 1972; Tann, J., *The Development of the Factory*, London, Cornmarket Press, 1970; Toynbee, A., *Lectures on the Industrial Revolution of the 18th Century in England*, London, Longman, 1906.

5. Mathias, *The First Industrial Nation*, p. 7.
6. Mantoux, *The Industrial Revolution in the Eighteenth Century*, p. 26.
7. Toynbee, *Lectures on the Industrial Revolution*, p. 38.
8. Ibid., p. 29.
9. Macdonald, S., 'Agricultural response to a changing market during the Napoleonic wars', *Economic History Review*, 33, 1980, pp. 59–71.
10. Mathias, *The First Industrial Nation*, p. 77.
11. Ibid., p. 79.
12. Thompson, S., 'The potential for and limitations of a shift from animal-based agriculture and food production in England and Wales', Ph.D. dissertation, University of Aston in Birmingham, 1979. See also Thompson, S. and Braun, E., 'Cropping for plant-based agriculture—a vegetarian scheme for England and Wales', *Food Policy*, May 1978, p. 197.
13. Toynbee, *Lectures on the Industrial Revolution*, p. 47.
14. Ibid., p. 49.
15. Ibid., p. 51.
16. Ibid., p. 53.
17. Ibid., p. 54.
18. Ibid., p. 57.
19. Mathias, *The First Industrial Nation*, p. 157.
20. Savery, T., *The Miner's Friend*, London, 1702, reprinted in Russell and Goodman (eds), *Science and the Rise of Technology since 1800*.
21. Mantoux, *The Industrial Revolution in the Eighteenth Century*, pp. 318–32.
22. Landes, *The Unbound Prometheus*, p. 103.
23. Mathias, *The First Industrial Nation*, p. 135.
24. Tunzelmann, von, G. N., *Steam Power and British Industrialization to 1860*, Oxford, Oxford University Press, 1978.
25. Landes, *The Unbound Prometheus*, p. 98.
26. Ibid., p. 104.
27. Ibid., p. 99.
28. Mathias, *The First Industrial Nation*, p. 135.
29. Ibid., pp. 282–3.
30. Mensch, G., *Das technologische Patt*, Frankfurt a.M., Fischer, 1977.
31. Mathias, *The First Industrial Nation*, p. 131.
32. Mantoux, *The Industrial Revolution in the Eighteenth Century*, p. 206.
33. Ashworth, *A Short History*, p. 8.
34. Tunzelmann, *Steam Power*, p. 198.
35. Landes, *The Unbound Prometheus*, p. 108.
36. Roesch, K., *3500 Jahre Stahl*, München, VDI–Verlag, 1979.
37. Ibid., p. 37.

38. See, for example, Checkland, *The Rise of Industrial Society*, p. 87.
39. Gross, N. T., *The Industrial Revolution in the Habsburg Monarchy 1750–1914*, London, The Fontana Economic History of Europe, vol. IV, ch. 5, 1972, pp. 20-1.
40. Ashworth, *A Short History*.
41. Ibid., pp. 62-3.

2   Technological Innovation

INTRODUCTION AND DEFINITIONS

Change is inherent in all human affairs, but technology has the relentless drive for innovation at its very core. For anything that is achieved can be achieved better, and any failure is a challenge for the future. Change and the search for new solutions are inherent in technology itself, but the driving force for innovation stems from the economic role which technology has assumed. It is the partnership between a technology seeking novelty and an economy seeking comparative advantage which creates a potent mix for innovation. Technology and economic interests can and do sometimes combine into deadly inertia; but the natural role of technology is to seek new avenues and new ventures and when this inclination finds an economic ally, innovation results.

Because old-established technology is hard to control—it is usually too deeply rooted and entrenched—new emerging technologies are the ones of greatest interest to those seeking to influence technology by policy measures. The support of technological innovation in general, the support of specific innovation projects, the regulatory intervention in the control of undesirable aspects of new technology—all these are rich fields for both student and practitioner of technology policy. These issues will be dealt with in later chapters, for the moment we shall merely describe the process of innovation.

Before embarking upon a discussion of the process of technological innovation, some basic definitions are necessary.[1] We define as a technological innovation an entirely new or technically substantially improved product or process which is offered for sale to potential users. This definition distinguishes carefully between an invention and an innovation, for an invention is merely an idea for or a prototype of a new product or process and does not become an innovation until it reaches the market. Most inventions never become innovations, they fall by the wayside on the long road from idea to marketable product.

We may further define as a successful innovation one which reaches sufficient sales to pay back the investment required to put it on the market and, for preference, achieves an actual net profit for the innovator. By contrast, an unsuccessful innovation is one which fails to repay its investment. This needs some qualification. The definition is adequate for goods or processes which compete on commercial markets and are judged by commercial criteria alone. There are some goods or processes, however, which may be of benefit to the public at large or to some specific groups, say people suffering from a specific disability, and yet cannot achieve adequate commercial sales. Such innovations may be judged worthy of public subsidy and may be highly successful on non-commercial criteria. Similarly, commercially successful innovations may have socially undesirable qualities. This is a difficult area of conflict between commercial and wider social interests to which we shall return on several occasions. For the moment we shall adopt the purely commercial criterion of success, i.e. whether or not the total cash flow attributable to the innovation is positive.

Innovations can be classified on any number of criteria. Four such classifications are shown in Table 2.1. The important dimensions

Table 2.1. Classifications of innovations according to four
different dimensions

Dimension	Type of innovation		
Object	Product	Process	Manufacturing
Type of market	Consumer	Small capital	Large capital
Degree of novelty	Radical		Incremental
Economic significance	Basic		Improvement

are the object of the innovation, the type of market it is aimed at, the degree of novelty and the eventual commercial significance. By 'object' is meant particularly whether the aim of the innovation is to achieve a new or considerably improved product, such as a pocket calculator, a disposable syringe, or a numerically controlled machine tool, or a new process, such as the manufacture of sugar from grain rather than from beet or cane. There is, however, a class of innovation which is important and is often subsumed under process innovations, but in reality is quite distinct from them. We define manufacturing innovation as 'a new way of producing an essentially established product by an essentially established process'.[2] Such innovation usually involves the use of machinery novel to the organisation and/or novel methods of machine or process control. Innovation may mean the first time use of a new technology in an organisation—new to the organisation—and need not mean new to the universe. Thus one firm's innovation may be another firm's sales: a robot installed in some factory is a typical major manufacturing innovation, yet it is also part of the diffusion process of robots and a marketing success for a robot manufacturer.

The type of market which the innovation might find does, of course, make all the difference. The marketing of a consumer product is a very different matter from the marketing of a small investment product, say an office typewriter, or from major capital equipment, say a glass bottle manufacturing machine. Often the categories overlap and 'large' or 'small' are relative terms, but on the whole these distinctions help to clarify the process and to realise what actions are necessary to improve the chances for successful innovation.

The essence of innovation may be described as the coincidence of a new technical possibility with a market opportunity. Some innovations occur more in response to a known market need— say the provision of car engines of frugal consumption and clean egestion; in others a new technology must capture a market for itself—say early pocket calculators which nobody knew they needed. In innovation jargon we speak of market pull and technology push to describe the extremes of an untidy range of intermediate situations.

The degree of novelty is an obvious criterion for classification. Generally we class an innovation as radical if it uses a new scientific principle and makes a real break with older technologies. The transistor or the ball point pen or nylon may serve as examples. Very often such breakthrough technologies arise, or become useful, only because of the conjunction of an old idea with a new context. An example is the planar process of manufacture of transistors, a vital step on the way towards integrated circuits, when a known process of printing technology was transferred to the fabrication of transistors.[3] People unfamiliar with the technology would hardly recognise the planar process as vital, indeed very few people outside the industry have ever heard of it. The question of what is or is not a breakthrough innovation is by no means trivial.

Major innovations often occur in clusters. This is not particularly surprising, as any complex and radically new technology comprises many components and many of these need to be new or redeveloped. Furthermore, the first of a new species usually has many weaknesses which subsequent innovations try to remedy. The classic example of the steam engine illustrates both points abundantly. The steam pump invented by Savery led to the first engine by Newcomen. The poor efficiency of Newcomen's engine was improved by Watt's separate condenser. Later inventions improved the efficiency further and made it possible to decrease the size of engines while increasing their output. Steam engines needed steam boilers, and these too were improved by numerous inventions. In particular Stephenson managed to obtain much more steam from a small mobile boiler by blowing steam through it, thus increasing its draught. The conversion of the to-and-fro movement of early steam engines to rotary motion required numerous additional inventions, such as Watt's planetary gears. The speed of engines had to be regulated and to this end Watt invented the centrifugal 'governor'. The transfer of Wilkinson's method for boring cannon to the manufacture of cylinders for steam engines proved a decisive step in the process of maturation of steam-engine technology.

It would appear that any successful radical innovation does in fact consist of a cluster of related innovations and this fact may provide a better definition of radical innovation than the conventional one, relating it to new scientific principles. We might define a radical innovation as 'a cluster of related innovations which together form a technology which differs considerably from previous technologies'. Unfortunately, both these definitions leave out some extremely ingenious innovations which cannot claim scientific novelty or large clusters; for example the cat's eye for marking roads, the zip fastener or the ball-point pen. The only consolation that can be offered these innovations is that they will appear under the category of basic innovations because of their economic importance. By contrast, an incremental innovation is one which offers a relatively small technical improvement without changing the nature of a technology. Thus a hovercraft is undoubtedly a radical innovation, but some improvement to simplify the manufacture of its skirt or to increase its stability would be an incremental innovation. It may be that the sum of all incremental innovations is more important than the sum of all radical innovations, but by their very nature they do not catch the eye.

Finally, there is the categorisation into economically basic and improvement innovations. It is tempting to equate the technically radical with the economically basic (and technological increment with economic improvement) and much confusion has arisen out of this temptation. The two systems of assessment are not equal and do not lead to the same result. While the degree of technical change can be assessed—perhaps by a vote of experts—on the basis of a technical description of the innovation or cluster of innovations, the economic significance can only be assessed with hindsight. Our previous examples may illustrate the point. While the hovercraft is undoubtedly technically radical, its economic impact is at best marginal.[4] On the other hand, the zip fastener, ball-point pen and cat's eye have created and satisfied large-scale market demands and have had considerable economic effects. The transistor certainly was scientifically radical, but only in conjunction with a large cluster of further innovations did it become an economically basic innovation which created a whole new industry.[5]

THE ECONOMIC SIGNIFICANCE OF INNOVATION

Technological innovation has been hailed as the saviour of stagnant economies, the life-raft for sinking firms and the determinant of long-term economic cycles. No doubt some of the claims are exaggerated, but no doubt also that innovation has considerable economic significance in the three areas mentioned.

On the microeconomic level, the level of the firm, innovation plays a major role as a determinant of competitiveness and growth. If a firm has little comparative advantage and suffers much disadvantage compared to its competitors, the firm will go under. Among the many factors which determine the position of the firm, technological innovation plays two decisive roles. Up-to-date products, preferably products slightly ahead but certainly not below the 'state of the art', are an inescapable primary condition of success. Given the same price and type of good, the quality, including the technical quality, of the product must remain comparable. Whenever the state of the art changes, the firm must be ready to go along or else it will lose its market share. We speak of defensive product innovation if a firm merely innovates to maintain its market share and of offensive product innovation if it attempts to get ahead of the pack by offering a technically more advanced product. In the extreme, the product may be radically new and the firm may attempt to create markets which did not previously exist.

The second role of technological innovation in determining the competitive position of the firm is in production technology. Unless a firm keeps its production technology up-to-date, which means that it engages either in process innovation when appropriate or in manufacturing innovation, it will suffer loss of competitiveness by losing out on factor productivity or on quality of production or both. The role of production technology and the forces which control it will be discussed in detail later, for the moment only its role as determinant of efficiency and quality of production is significant. In theory, and to some extent in practice, an equal product can be achieved with different proportions of capital and labour inputs. The function relating these factors of production is known as a production function.[6] If the technology changes, then the production function changes too and normally the same product can be achieved with fewer inputs, therefore with higher productivity. Higher productivity will eventually, in a competitive situation, lead to lower prices and a firm with low productivity will not survive unless it is protected from some aspect of competition. Other things being equal, a firm with poor production technology will not only do badly because of low labour productivity, but also on other counts and thus manufacturing innovation may be aimed at improved quality, fewer rejects, lower consumption of energy or materials, lower stocks of components, greater safety for workers, or less environmental pollution. As all these things may help the firm to gain comparative advantage or comply with external pressures, they too may be important factors in the well-being of the firm.

If we shift our viewpoint from that of the firm to that of the economy at large, technological innovation plays several significant roles. In the same way as it determines the competitive position of individual firms, so, integrated over the firms comprising an industry, it determines the competitive position of whole industrial sectors in national or, more commonly, international competition. The rise and fall of industrial sectors in different countries is too familiar a spectacle to need specific examples. Single-cause analysis of a complex phenomenon, such as the relative rise or decline of an industrial sector in a given country, tempts by its simplicity. The temptation should be resisted, for the simplicity is gained merely by gross falsification of reality. The rise or decline of an industry may have—and usually has—many causes, but among them technology takes pride of place. A sector may decline because its technology as a whole is overtaken (for example sail-making when steam ships became common) or it may decline nationally if the international competition operates with superior technology or under more favourable conditions.

It could be argued that the decline of an industrial sector, or even of the whole of industry in a particular country, are unimportant. If things can be produced more efficiently elsewhere, then everybody gains by buying from wherever the goods are manufactured with the highest efficiency. Such is the nature of the international division of labour. Which is all very well, except that industry serves two vital purposes other than the direct satisfaction of material demands: it plays an essential role in keeping at least a neutral balance of international trade and it provides employment. Of course there is trade in services, agricultural products and raw materials as well as in industrial products, but the very nature of an industrial nation, with its huge markets for industrial products, requires some domestic manufacture of such products, as this alone can assure the long-term balance of foreign payments. Although industry now provides only about one-third of total employment in advanced countries and much employment is provided in services, the employment of one-third of the working population is not insignificant. Moreover, much service employment depends directly on industrial activity, as services to industry form a major component of the service sector. The total well-being of an advanced industrial economy depends quite critically upon the well-being of its manufacturing industry, and although the sectoral composition may change with time, the uncompensated decline of any sector spells trouble for the national economy. Compensation may entail a degree of shift from manufacture toward services, including services provided for final consumption, but because of the need for balance in foreign trade this shift is unlikely to go very far, unless consumer spending shifts away from manufactured goods towards services and total foreign trade is thus reduced or significantly altered.

We may disregard the frequent statement that only industry creates wealth, for it implies the assumption that wealth consists of manufactured goods only, a statement which is as false as it is crude. Yet we must accept the argument that a healthy manufacturing industry is essential to the economic health of an industrial nation—almost a matter of definition—and thus the argument about the importance of technological innovation in the affairs of a firm can be directly transferred to whole industrial sectors. For a sector is only as healthy as the firms comprising it, and if technological innovation is vital to the firms it is vital to the sector and manufacturing industry as a whole.

So far we have concentrated on technological innovation as an important, though by no means sole, determinant of the successful survival of manufacturing industry. Yet industries and whole economies wish to do more than survive; they wish to flourish and to grow. Thus 'innovate or die' is only part of the battle cry, the full cry would include some phrase such as 'innovate to grow' or 'growth through innovation'. Innovation makes growth possible because new technology satisfies previously unsatisfied or even unarticulated demands. It does this on the one hand by creating new products and on the other hand by increasing productivity in the economy, thus making more resources available for consumption.

Growth in GNP (gross national product, which is a measure of total economic activity in a country) can be caused by increased inputs of the main factors of production (capital and labour) or by more efficient use of the factors of production. One of the means of better utilisation is by technological innovation. Several economists have used a variety of methods in attempts to measure the contribution of innovation to growth.[7] Denison's analysis of the contribution of various factors to economic growth yielded 23 per cent for the United States and 33 per cent for Britain as the contribution of new knowledge, while an analysis for the United States by Solow gave 87 per cent. Although the numerical results of such analyses have to be treated with caution, there can be little doubt that technological innovation is a major contributing factor to economic growth. As an example, the analysis by Denison, as reported by Green and Morphet, is reproduced in Table 2.2.[8]

As far as the economist is concerned, all demands or needs are equal. Anything that people are willing to pay for, any item of final consumption—bread, a ball-point pen with a built-in watch, beauty treatment, tax advice—they all are needs and the transactions enter GNP. To the philosopher some needs may seem more equal than others, but the debate very rapidly enters treacherous ground signposted by will-o'-the-wisps: personal freedom, value judgements, allocation of utility. We shall not avoid these issues but must defer them to the policy debates where they properly belong. For the moment we must accept that technological innovation is an essential part of the competitive life of firms and economies in any economic system which contains competitive elements, and therefore in virtually all systems in operation today. We must also accept that technological innovation is closely associated with economic growth—for better or worse.

Much discussion has centred on the issue whether, and how, technology determines long-term cyclical fluctuations in the world economy, the so-called Kondratiev cycles. Freeman, Mensch and others assume the upswings in the roughly 40–60 year cycles to be initiated by the introduction of basic innovations, which then give way to improvement innovations, merely improving productivity without giving rise to new opportunities and thus leading to the downswing in the cycle. The major technologies of the industrial revolution—textile machinery, steel production and the steam engine—are assumed to have caused the first cycle of prosperity in 1787–1800, recession in 1801–13, depression in 1814–27 and recovery in 1828–42. The second Kondratiev cycle is associated with the railways and covers the period 1843–97, while the third cycle is ascribed to electricity and the chemical industry. Our current

Table 2.2. Summary of Denison's analysis of UK growth rate 1950–62

UK growth rate	Breaks down into:	Breaks down into:
2.29% annual growth between 1950 and 1962	1.11 owing to increased inputs	0.6 owing to increased inputs of labour (0.29 of which owing to education)
		0.51 owing to increased inputs of capital
		1.11 subtotal
	1.18 therefore owing to increased output per unit input	0.03 owing to more efficient use of old knowledge
		0.12 owing to improved allocation of resources
		0.36 owing to economies of scale
		─0.09 owing to irregularities in demand
		0.42 subtotal
		0.76 therefore the residual, assumed owing to advances in knowledge = 33%
		1.18 total

Source: After Green, K. and Morphet, C., *Research and Technology as Economic Activities*, London, Butterworths, 1977, p. 18.

economic difficulties are said to mark the downswing in the fourth Kondratiev cycle.[9]

It was Schumpeter who first introduced the notion of technological entrepreneurship as a causative factor in economic upswings and the Schumpeterian view has influenced much thinking about the role of technology in the economy and particularly about technological causation of economic cycles. Three features of technology stand out which may cause cyclical economic change. First, technological innovations open up new areas of economic activity in that they satisfy new needs and provide new investment opportunities. Second, products undergo a complex product cycle from their initial introduction (innovation) to their routine manufacture and possible demise. Third, all goods, and particularly investment goods, have a limited life and require eventual replacement.

When a basic innovation first appears on the market—a product such as the motor car or the electric motor—it opens up a multitude of possibilities for new economic activity, quite apart from the production of the motor car itself. Such basic innovations are extreme examples; products which truly open new economic vistas are extraordinarily rare. The search for such innovations is equivalent to the alchemist's search for gold. But even if the new product does not inaugurate an entire new range of activities, even if the product is a modest one, it still offers new investment opportunities and satisfies new needs—thus enlarging total economic activity and causing growth.

Usually the new product requires highly skilled workers for its development and initial manufacture and usually the production is not immediately efficient. This is particularly so in the case of high technology products.

Without being able to define high technology products we can point out a few of their hallmarks: considerable research and development effort; the need for sophisticated and highly qualified personnel in development, design and initial manufacture; initial high cost and price per unit; use of sophisticated scientific measurement and control instruments in manufacture; complex and varied parts, materials and techniques.

Quite apart from patent protection, the innovator in high technology enjoys an initial near monopoly position caused by the inability of the competition to imitate the complex processes. Gradually, the manufacture of the high technology product becomes routine. The processes are improved, manufacturers of ancillary equipment offer their wares on the market, thus enabling later entrants to buy machinery which the innovator had to develop. As the new product becomes more ordinary, it requires fewer skilled workers to produce it and as the production builds up the output per worker increases. The initial monopoly is eroded first by competition from enterprises within the advanced countries and eventually from all over the world. As competition increases, so prices and profits fall and it often turns out that an overinvestment has taken place, so that incremental funds lead to diminishing returns. This is exacerbated by problems of market saturation. The production of transistors or of colour television sets and video recorders demonstrate all the features described. Hence there is pressure for ever new products, new investment opportunities, and new escape routes from relentless competition.

Finally, investment goods and all products have a finite life. Even the best machine wears out and needs to be replaced. This finite life leads to a natural investment cycle, different for each product. This is unavoidable, but questions arise if replacement occurs because of obsolescence rather than of wear and tear. From the point of view of conservation of resources, machines should last as long as possible and should be allowed to serve to the end of their natural life. Relentless technological progress tends to force them into early retirement.

Technology itself thus has several cyclic factors in its very nature and it is not surprising that technology has been looked at to provide explanations for cyclic behaviour of the economy. However, technology is not the only economic force with cyclic tendencies and although it must be regarded as a contributor to cyclic behaviour, it is only one of many. The recent oil shock to the world economy, monetary policies, wars, and a variety of inflexibilities spring to mind as examples of non-technical influences on economic cycles.

Economies fluctuate, for technological as well as non-technological reasons, in irregular cycles. Mathematical analysis describes complex cyclic behaviour as a series of simple sine waves of different frequencies. Thus any complex fluctuations will have long-term components and it is doubtful whether a separate cause needs to be found for these. They may simply be regarded as bunching of separate short-term cyclic forces. Once such bunching occurs, it has the unfortunate habit of self-reinforcement, no doubt often exacerbated by incorrect political remedies.

In a recent analysis of long-term economic activity, from 1850 to 1979, Haustein and Neuwirth found they could fit the data with several cycles, ranging in duration from seven to fifty years. They attempted to find correlations between industrial production, innovations, patents and energy consumption. The results yield only rather obscure correlations, but it must be borne in mind that any list or index of inventions is bound to be arbitrary and incomplete and therefore any such analysis is fraught with difficulties. The only objectively measurable quantities are numbers of patents, energy consumption and total industrial production, and the correlations found between these make very limited sense in that they show a correlation between US patents and industrial production in the seven, thirteen and fifty year cycles.[10]

Much discussion has centred on the question of whether inventive activity leads to upswings in the economy or whether upswings in the economy cause increased innovative activity.[11] It must remain as unanswered as any chicken and egg question: growth and innovation feed on each other and either can take a localised lead, but on the whole they go hand in hand. New technology creates new investment opportunities, and the availability of capital helps to create new technology.

THE INNOVATION PROCESS

As a very rough approximation, product innovation may be schematically described as a sequence of events shown in Figure 2.1. The details vary enormously according to types of technology, firm, and innovation. Process innovation fits into a somewhat similar scheme, although it may end at the 'prototype' stage to be sold as a process, or it may trigger off a series of product innovations, required to implement the new process. In reality, the stages overlap with each other and there is continuous change and feedback within the innovation process.

We shall now attempt to describe innovation in a general way, which should capture the essence of product, process and manufacturing innovation and describe the undertaking of innovation as a sequence of phases, each with its own constellations necessary to allow progression into the next phase. As in all generalisations we shall lose much detail, but hopefully we shall gain a framework which will help to make sense of detail in otherwise rather unconnected case studies of innovations. Innovation proceeds through several phases with complicated feedback between them. The phases describe a logical sequence, temporally they may overlap considerably.

Before describing the phases of the actual innovation process, we must describe the firm involved in the attempt to innovate and the world it lives in. This description will include size of the firm, types and ranges of technologies used, markets in which it operates, market position, general business environment, technical, managerial and research capabilities available. We call this phase the zeroth phase, for it describes the situation and capabilities without entering upon a description of the innovation itself. A description of the family before conception of the infant, as it were; although the firm must remain aware of changes in its own

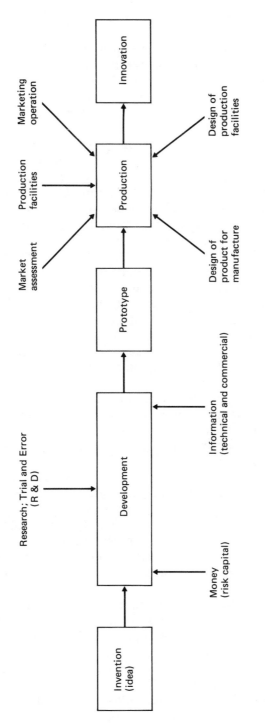

Fig. 2.1. Product innovation as a sequence of events. *Source:* E. Braun, 'Government policies for the stimulation of technological innovation', IIASA Working Paper, WP-80-10, January 1980, pp. 4–5.

organisation and the world surrounding it throughout all phases of the innovation. Even at this stage we are describing a constellation of forces and circumstances, in general keeping with a constellation view of causation. The net result of the zeroth phase should be an appraisal of the range of the possible and the necessary, and any firm will ignore this phase at its peril.

The first phase of an innovation consists essentially of the identification of a need–possibility relationship. In product or process innovation the 'need' may be a known market demand (market pull) or it may be the hope of creating a new need (technology push). The 'possibility' is always technological; a new technical path is envisaged, a new way of doing something. The new possibility often occurs because of new research findings or because of the rise of a new technology in a different field. Cross-fertilisation across technical and scientific disciplines or sub-disciplines is one of the most common modes of creating new technical solutions.

In manufacturing innovation the first phase consists of the identification of a weak link in the manufacturing system—the need—and of a possible solution to it. The weakness may only exist because of the rise of new technical possibilities, for often weaknesses become such only in the face of superior solutions.[12]

Each phase of an innovation needs 'actors' to make it proceed. Several roles have been described, for example by Freeman. The most important among them is the role of product champion, defined as 'any individual who made a decisive contribution to the innovation by actively and enthusiastically promoting its progress through critical stages.'[13]

At the end of the first phase a well considered idea for an innovation emerges, based on an expected confluence of market need and technical possibility. A group of actors who have progressed the idea to this point, taking into account their knowledge of the zeroth phase, should now be prepared to proceed to the next phase. The driving force will be the product champion, who may be an individual or a small group of people with adequate power within the organisation attempting the innovation.

The second phase of the innovation is the solution/development phase. In product innovation this is the period when the idea is developed into a working prototype, ready for production, and an appraisal is made of market potential and marketing problems. For example, in the case of a pharmaceutical innovation, at least all the prescribed toxicity tests will have been carried out and initial clinical trials, though perhaps not full-scale clinical appraisal. For a process innovation much the same applies. The process will have been taken to a stage where it is known to work, at least in a pilot plant, and where its technical/economic performance characteristics and its hardware requirements are fully known. For a manufacturing innovation, all special development work and testing of alternative solutions falls into this phase, so that by the end of the phase management will know exactly how it wishes to solve the production problem identified in phase one.

Phase two of an innovation may involve not only a great deal of work and large expenditure of time and money, it also requires the right constellations of forces and circumstances to prevent the foundering of the innovation on any number of treacherous obstacles. For example, the problems arising in development may need further research which must yield useful results at the right time. Technologies and materials which are required for the new product/process must be available, as must specialist services or specialist manufacturing techniques. An example may illustrate the point, but should not detract from the enormous range of possible problems.

The big breakthrough that made the innovation of a supersonic passenger airliner possible was the scientific/technical realisation that a special design of wing, the so-called thin swept wing, could reduce aerodynamic drag at high speeds sufficiently to make the aircraft possible, yet would remain stable at landing and take-off speeds.[14] Thus the confluence of possibility and need in the first phase consisted of the technical feasibility of building a supersonic airliner and the perceived need of the flying public to travel at ever greater speeds. The historical trend toward increasing speed of travel was regarded as convincing proof that the trend would continue if it could. The total constellation of factors emerging from the zeroth and first phases of the Concorde story was complex, but in summary the British aircraft industry felt able to produce the aicraft, believed that a market for it would exist, and convinced the British government to foot the bill in order to keep the civil aircraft industry alive. A further dimension was added when Concorde became a test-bed not only of aerodynamics but of the political dynamics of Anglo-French cooperation.

The second phase of the innovation required a vast research, development, design and re-design programme and also developed into a political struggle. Technically, it was necessary to find an engine to give the necessary thrust; to design and construct air intakes of variable geometry to allow the engine to operate within the full range of speeds; to design a fuel pumping system to shift the centre of gravity of the aircraft as it moved from subsonic to supersonic speed; to design a special heated windscreen; a variable geometry nose-cone; and dozens of other items all depending on the latest state of the art in servo-mechanisms and control theory, computing, aerodynamics and materials research. The project also depended on a vast network of specialist suppliers, on the availability of adequate wind-tunnel facilities, a team of first-rate experts on every aspect of aeronautical engineering, and much else besides.

Some of what has been nonchalantly called 'much else besides' caused the eventual commercial downfall of the technically brilliant aircraft. The design had to be negotiated and re-negotiated with the airlines, and too many changes became necessary. The fit between the technically feasible and the commercially desirable proved somewhat problematic. Unforeseen major technical, commercial and political problems emerged which proved inauspicious: the supersonic boom, noise during landing and take-off, and very high fuel consumption per passenger mile compared, particularly, with the new generation of wide-bodied air-liners with their fan-jet engines (a factor of importance especially after 1973). Thus some factors and forces in the constellation proved auspicious and could be controlled by the programme managers; others proved inauspicious and beyond their control. Should they have foreseen what they could not control, and should they have called off the attempt when the climate became too harsh? In so major a project the development time is so long that the world changes and the appraisals of the zeroth and first phases have to be repeated many times. Perhaps Concorde was a good idea when the project started, but the idea became addled with changed circumstances and only politics prevented a timely review and cancellation.[15]

Different innovations encounter and require different constellations of circumstances during the second phase of the innovation process. Many falter, some survive. At the end of the second phase, those that survive will be in advanced prototype form in the case of a new product; a working pilot system in the case of a new process and a tested and chosen solution in the case of a manufacturing innovation.

The third phase of the innovation is the implementation phase. For a product innovation the major requirement is the provision of manufacturing facilities to make the product, which emerged from the second phase as a prototype, in commercial quantities. Temporally some of the phases may overlap and manufacturing facilities may be designed and built while prototype development work is in progress. Our phases are meant to describe a logical rather than a temporal sequence. It is vital, of course, that prototype development should consider requirements of 'manufacturability'. With some products, and microelectronic circuits are a prime example, the ability to make the item exerts a decisive influence on the product itself.[16]

At the same time as manufacturing facilities are prepared, marketing must also gain a new urgency. While market considerations are vital to the very idea for an innovation and market coupling is important throughout development work, once manufacturing preparations are being made, thoughts on how to sell the product must come pretty near the top of priorities.

In process innovation, the implementation phase means moving from pilot plant to full-scale operation of the new process. Often, this procedure is similar to the one in product innovation, for the full scale plant needs to be built and its requirements may modify the initial process. While in the pilot plant some controls might still have been manual and it may now be necessary to develop full control programmes and perhaps even sensors for some process variables.

In both product and process implementation a major role is played by organisational factors. It is necessary to recruit and train suitable personnel and to set up a suitable management and administrative structure. All kinds of new services and supplies need to be arranged.

In manufacturing innovation, the main thrust of the implementation phase is organisational. Complex questions of the allocation of tasks, reallocation of skill requirements and wages settlements have to be solved and often new hierarchical structures will emerge.[17]

The fourth and final phase of the innovation is the consolidation phase. At this stage continuous improvements are made to the technology itself—whether to the product, the process or manufacturing procedures—and to the organisational and marketing arrangements.

A vital concept during this phase is the 'learning curve'. In a large number of small steps the manufacturing process, the organisation and the product are gradually improved. The result is that as the production is stepped up, so the gradual improvements as well as economies of scale come into operation and unit costs fall, sometimes dramatically. The learning curve is really separate from economies of scale, but in the usual way of plotting costs against production run the two factors become as one. The learning curve is a result of conscious or unconscious improvements in skills, methods, materials, design, organisation, marketing and management. The very words 'learning curve' express clearly what happens—

people learn by doing and draw lessons from their experience.[18] Economies of scale, on the other hand, result from a fuller utilisation of machinery and labour as the rate of production approaches the capacity of the production line, out of more favourable trading terms for larger quantities and many similar factors.

Both the learning process and economies of scale are important ingredients of innovation. New products normally have to be subsidised not only to the point of production, but long after their initial introduction to the market. The true food of innovation is hope and only during the consolidation phase does it become clear whether the hope will be fulfilled. Figure 2.2 shows a schematic learning curve and describes the cash flow for the innovating firm, which is one of the crucial parameters in innovation.

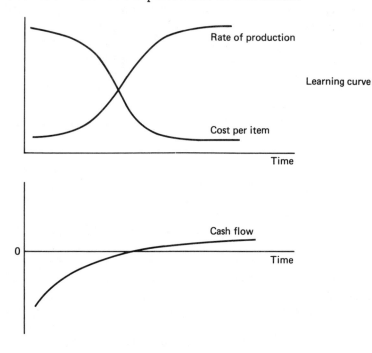

Fig. 2.2. Schematic learning curve and cash flow.

Toward the end of the fourth phase the innovation becomes more of a routine operation of the firm. As a result of the innovation the firm may have changed in many ways and the fourth phase thus feeds back into the zeroth phase. The circle is closed, the transformed firm, living in a transformed world, is ready to embark upon the next innovation. In a large firm, of course, several such cycles will be running simultaneously and there will be innovations at different stages at all times. Table 2.3 summarises the phases of the innovation process as described above.

Much empirical research has been done on the question of what factors determine success or failure in innovation and on the related question of how firms go about innovating. Doing less than justice to the many extensive empirical studies, we may summarise their findings into four major points.[19]

The role of committed and competent people—especially but not only the innovation champion—is crucial. In large firms or small, the innovation champion must be able to convince the controllers of the purse strings to devote sufficient resources to the particular project, perhaps at the expense of other pet ideas. Even more important is the ability of the project management to coordinate all the manifold activities which are necessary to bring technical and commercial success. The management of the spectacular space exploits by the US National Aeronautics and Space Administration (NASA) has become legendary, but many a humble innovation project requires the same kind of coordination, perseverance, juggling and sheer hard work. The auspicious constellations required throughout all the phases of an innovation project do not just happen—they have to be brought about by deliberate effort. Luck is, of course, an essential ingredient of success; but in innovation at least luck helps only those who could very nearly succeed without it.

The second major empirical finding might be summed up as good communications with customers and with relevant expertise. The results of project SAPPHO particularly stressed good liaison with potential users and support for them.[20] Although this particular finding holds specifically for process innovations in the chemical industry and for scientific instruments, i.e. cases where the innovator deals with a circumscribed and expert set of potential users, it holds in some form for many other innovations. It was found, for example, that one of the crucial factors which determine success or failure in the application of industrial robots is the support available from the robot manufacturer to the user. This support includes detailed advice on the engineering and logistic problems of the particular applications, training of operatives and maintenance personnel, and provision of effective back-up maintenance services.[21] Perhaps this is not surprising, as a manufacturing innovation, such as the introduction of robots, is a capital-goods innovation from the point of view of the robot manufacturer. Yet even the introduction of microprocessors would not have been the commercial success it eventually became, had not the innovators, more particularly INTEL,

Table 2.3. Phases of innovation

Zeroth phase: boundary conditions	First phase: idea	Second phase: development	Third phase: implementation	Fourth phase: consolidation
Description of the firm and its immediate environment	Identification of a possibility —need constellation	Development of product or process or solution of manufacturing weakness	Provision of manufacturing facilities for product	Learning and improvement
Appraisal of the capabilities and limitations of the firm	Emergence of idea for product or process or for remedy of manufacturing weakness	Research, development and design effort; location of supplies; market coupling	Full-scale operation for process	Sales, economies of scale
		Emergence of prototype product or pilot plant process or solution to manufacturing problem	Organisational arrangements for manufacturing innovation	Organisational finalisation
			Marketing and sales	Feedback to zeroth phase

very rapidly branched out into alien territory and begun to supply software support and so-called development systems which made it easier for their customers to use the new devices.[22]

That good communications between the innovator and those with the relevant expertise were found to be an important factor in success is not very surprising either. After all, unless up-to-date knowledge of the relevant areas of science, technology and market needs is available, it is unlikely that the firm will produce a competitive product or process. On the other hand, it is unlikely that all necessary knowledge will be available in-house, so that communication with external expertise is vital. The key to success is selective intake of information by people capable of both selecting and using such knowledge.

All innovators regard patents as of great importance and use them extensively. The role of patents varies in different industries and has generally moved away from the original role of granting the owner of a patent a monopoly, albeit a temporary one. In modern use patents do give a degree of protection, but generally licences are granted to other manufacturers for the use of the patent and licence fees can be an important source of income. In some industries, and the semiconductor industry is a prime example, patents have become a trading commodity. Firms enter into mutual licensing agreements and the number and quality of patents held by any given firm will determine not only its 'balance of trade in licence fees', but also its leverage in obtaining licences from reluctant licensors. A strong patent position can influence the standing of a firm *vis-à-vis* investors, creditors, governments and take-over predators.

As a rule, internally produced innovations which are not detectable in the product, such as new processes or manufacturing procedures, are protected by company secrets rather than by patents. Protection by secrecy is preferred to reliance on the law.

Special problems of protection of property rights arise with some modern technologies, where neither secrecy nor patents provide adequate cover. Software protection is one such area, where patents are inapplicable and copyright is of limited practical utility. In such cases, protection is often sought by technical tricks—the technological fix—such as electronic locking systems built into the software.[23]

The feature which distinguishes most clearly between success and failure in attempted innovations is the emergence of a product or process which suits consumer needs and which functions well without major modifications. The innovation must enter the crucial consolidation phase fully developed and fully suited to its use and user, otherwise it is likely to fail or falter at this late stage.

After long periods of investment into an innovation, firms are strongly tempted to try and reverse the negative cash flow as early as possible, and sometimes start marketing a product which is not fully developed and free of defects. Sometimes also, the innovator has inadequate knowledge of the applications to which the innovation will be put. Both these phenomena occurred in the development of industrial robots. Early robots proved unreliable and gave the technology a bad name, from which it took a long time to recover. In the early days also robots were used in inappropriate applications or in clumsy ways.[24] Robots survived these early trials, but many an innovation or innovator did not.

The final conclusion of project SAPPHO was that successful innovation required, particularly in the chemical industry, strong in-house professional R & D. The general validity of this conclusion must be in some doubt, as all the firms in the chemical industry had such a capability and it was thus not a distinguishing factor between success and failure. On the other hand, in the instrument industry there were 'cases of attempted innovation without such a structure. Most of these were designs for a new product brought from an outside environment.'[25] Project SAPPHO dealt with the chemical industry, an early example of a science-based industry which remains highly research intensive, and the instruments industry, which is almost by definition closely linked with scientific research. Indeed the coupling between professional scientific research and technological innovation has become very strong, so strong that the two are often confused. Perhaps we may at this point digress somewhat from the discussion of innovation to clarify the relationship between science and technology in general and technological innovation in particular.

THE SCIENCE-TECHNOLOGY BONDS

Science and technology have, since the time of Galileo and Newton, developed a multiple interdependence. The precise relationship varies not only from science to science, technology to technology, and innovation to innovation, but also from phase to phase of any innovation. Three different pathways of science–technology interactions and links appear of particular importance.[26]

The first link we might call the instrument bond. The model, shown in Figure 2.3, postulates a bond between science and technology formed by shared development and shared use of scientific instruments. There is a constant interplay between the development of 'scientific instruments', defined here in the broad sense of 'equipment

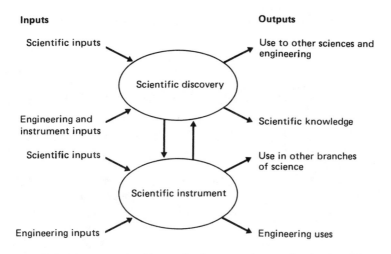

Fig. 2.3. The instrument mode of interaction between science and technology (the Instrument Bond).

used for purposes of conducting scientific experiments', and both science and engineering. We find even in this mode that the relatively simple model of a scientific discovery leading to a new instrument leading to further discovery is inadequate. Instruments and discoveries require a range of propitious circumstances and are, in a sense, a result of the current state of the scientific–technical system. It is also important to bear in mind that science is by no means the sole user of scientific instruments. In fact the equipment used for the purpose of scientific experiments may be used in many engineering applications, such as product development, quality control or even manufacturing processes. The very concept of 'scientific instruments' proves inadequate as the science–technology relationship becomes increasingly incestuous.

The legendary scientific inventor is more likely to invent an instrument than anything else, and an instrument might be adequately engineered and profitably marketed on a relatively small scale. Thus the instrument market offers considerable scope for scientifically trained entrepreneurs. The bond can only function if there are good open communications between scientific and technical activities. These can be secured by instrument makers keeping in constant touch with scientists in their relevant fields, both in order to sell their instruments and to gather ideas for new ones. A picture of a scientific community in close active contact with an entrepreneurial group of suppliers of scientific instruments seems to be the most appropriate mode of catering for the instrument bond between science and technology and thus for innovation in the instrument industry.

The second mode of interaction between science and technology is that leading to product development or product innovation and is closest to the traditional linear view of 'science invents— technology develops'. A possible model for this interaction is shown in Figure 2.4. The bond between science and technology is formed by the common interest of product or process innovation—they share, as it were, the same concern and the bond between them is deliberate and organised.

It is vital for success that the scientific and engineering activities should be closely linked. Scientists far removed from development and production neither can nor will contribute significantly to the innovation process. The bond is a cooperative one and that means that engineers and scientists contribute to all stages according to their abilities and according to the varying needs of the innovation as it proceeds through its phases. As the point can hardly be over-stressed, it might as well be said here again that commercial and marketing considerations do form the third partner in the science-engineering-marketing troika. The 'lead horse', as it were, varies from phase to phase, but in essence all three 'horses' are equal. It is this mode of coupling which generally requires a good in-house R & D capacity of the innovating firm.

There is no doubt that the greatest direct stimulus that science can give to innovation occurs when the science is practised in an atmosphere of technological enterprise with innovation in mind. Research oriented towards technical innovative goals is more likely to achieve such goals than pure serendipity can hope for, provided the quality and level of research are appropriate and adequate and the coupling between the two activities is close.

At different stages of innovation science assumes different roles. Even in semiconductor electronics, an innovation more heavily dependent upon science than almost any other, the direct role of science diminished in later stages of the innovation. While up to the mid or late fifties science, and more especially physics, was dominant in the production of new devices, production technology became the dominant partner from the sixties. Today, the dominant issue in the semiconductor industry is that of chip architecture, a matter within the orbits of art, magic and computer engineering. At no stage, however, were engineering or commercial considerations far from the centre of activity even in this science based innovation.

The third mode of interaction between science and technology is probably the most important, but also the most elusive. The

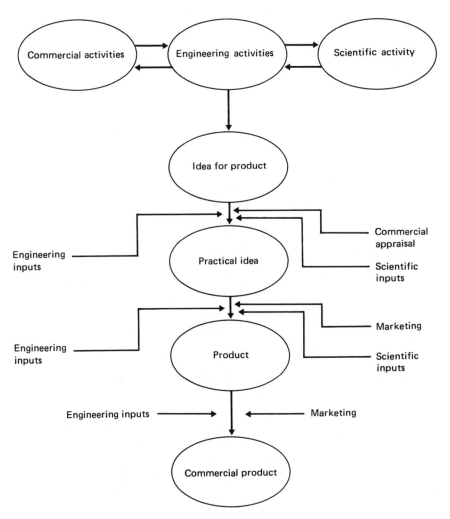

Fig. 2.4. Model of major contribution from science to product or process innovation (the Innovation Bond).

bond between the partners is formed by the joint creation and joint use of a pool of formal and informal knowledge. Science provides a reservoir of information, experimental and theoretical, on which engineering can and does draw as and when required. Similarly, engineering provides tools, materials, equipment and knowledge which science can draw upon.

We may look to semiconductor electronics for numerous examples of this kind of relationship. To use an up to the minute example, modern mask making equipment for integrated circuits uses an electron beam, a laser interferometer and computer control. The latter does, of course, depend heavily upon integrated circuits. Both electron beams and lasers are characteristic results of the 'instrument bond' between science and technology. Both were picked out of the common pool of knowledge by semiconductor device engineers attempting to provide ever higher densities of components on a silicon chip.

In addition to using a common inventory of materials, equipment and formal knowledge, science and engineering also use a common pool of skills, which we may regard as informal knowledge. It takes highly skilled technicians—in the broadest sense—to design and operate scientific instruments, manufacturing equipment and modern products. The training required is obtained within the scientific-technical activity and its specialised training functions, from apprenticeships to university post-graduate degrees in science and engineering. Scientific and engineering training both provide skilled people and scientific–technical attitudes which form a common reservoir of manpower for the modern scientific–technical enterprise.[27]

The 'common pool bond' is the least understood, yet probably the most important relationship between science and technology. It is difficult to ensure maximum efficiency of a process which inherently relies so much on chance and serendipity, but all the recommendations about 'open communications to the scientific world', addressed to innovators, aim at the strengthening of this bond. Open communications may require firms to entrust somebody with the role of 'gate keeper', the person who scans developments and brings promising ones to the attention of those within the firm who might profit from the knowledge. The 'common pool bond' requires a whole host of formal and informal networks of communications and a deal of mobility of people. The vast array of very problematic arrangements for the dissemination of scientific and technical information are all designed to improve the operation of the common pool bond. A major aim of technical and scientific education is to teach people where to find knowledge, how to recognise and plug gaps in their information; we might say they are taught how to fish in the pool of knowledge.

Much innovative activity consists of small incremental steps or of organisational arrangements for the assimilation of new technology. In these cases—and they form a large class of highly important innovations—the science-technology links may not be invoked. In all innovations involving high technology, almost by definition one or more forms of the science–technology bond will play a major role and it is important that the innovator should understand them in order to ease their operation.

THE DIFFUSION OF INNOVATIONS

The final fate of an innovation becomes apparent only in the consolidation phase. Shifting the viewpoint from preoccupation with the development of the innovation itself to its reception in the world—looking out from the inside—the consolidation phase becomes coincidental with what is often referred to in the literature as diffusion. The new process or product diffuses, in analogy with the diffusion of salt from a crystal of rocksalt immersed in fresh water. The analogy is a poor one, as the diffusion of an innovation is determined by a multitude of purchasing decisions whereas water does not have the option of not accepting the salt. Nevertheless, the innovator is pushing his wares from their 'point of origin' and they gradually spread, over time, through their potential markets— or not, as the case may be. Viewed mathematically, the amount of rocksalt dissolved and diffused into solution over time looks similar to the market penetration of a new product, with the crucial difference that market penetration is not certain and is subject to choice, resistance and inhomogeneity, not to mention the fact that the product may undergo changes—while the diffusion of salt in water is involuntary and homogeneous.

The first purchases of a new technology are pioneering acts. The first purchaser tends to have to pay a high price and often has to contend with 'teething troubles'. The decision to purchase will be governed, in economic terms, by the purchaser's perception of the utility of the new product or process. The purchaser will decide to

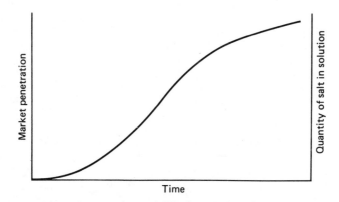

Fig. 2.5. Market penetration over time and diffusion of salt analogy.

buy if the expected advantages of the new technology are worth, to him, the price asked for it. In the case of capital goods, the decision can be based on estimates of savings, improvements in productivity and other tangible advantages. Yet almost invariably the first estimates, based on guesswork rather than solid experience, are unreliable and the decision contains an irrational element. Purchasing decisions follow the common pattern of mixed motivation and contain some elements of cool calculation, some of expectations of advantages which are difficult to quantify and some of one-upmanship and other wishful thinking.

There appears to be a degree of trade-off between the quality of a product and the need which it is designed to satisfy. If the need is great, then even a crude and unsatisfactory technology will be purchased to satisfy it. Examples abound in medical and dental technologies and many other areas. On the other hand, if a technology offers a great deal at low cost, then it will create a need for itself.

Once the first pioneering sales have been made, a great deal will depend on the experiences of the first purchasers. Knowledge of such experiences spreads by a variety of channels and there is nothing like satisfied customers to enhance sales, and nothing more likely to kill a product than dissatisfied ones. Thus the pioneers of diffusion are crucial in many ways. In the case of the introduction of new manufacturing techniques we deem the early purchasers to be innovators in their own right to such an extent that we have used the term manufacturing innovation for the process.

The pattern of diffusion differs greatly from product to product and process to process. In microelectronics, for example, we distinguish three distinct pathways of diffusion: the ready adoption of independent equipment, sporadic and delayed adoption, and exponential adoption.[28] The first of these categories relates to items such as digital watches or pocket calculators or video games. Such items are purchased by individual decision, can be used without major organisational changes and the impact of all the purchases is limited to a small sector of the economy. The second category, sporadic and delayed adoption, is concerned with either capital equipment, such as telephone exchanges or robots, or with equipment to be incorporated in some other major product, such as automobile electronics. In these cases, not only are the decisions to incorporate microelectronics complex and corporate, but they also depend upon large investment cycles and, when taken, affect large sectors of the economy. The third category, exponential adoption, means 'diffusion of innovation at a rate related to previous diffusion and adoption unconstrained by any finite market'.[29] The prime example is the computer and all that relates to it. So far, no limit is discernible in the market for machines to manipulate and store information and as more and more information is generated, so more and more machines are required to manipulate and store it. It is true, of course, that 'trees do not reach the sky', but the computer tree seems to have a long way to grow and, even more important, its growth feeds on itself.

Arguably, microelectronics and the computer could only have grown together—microelectronics providing the technical capability and computers capturing markets. This form of symbiotic relationship is quite common and is an aspect of the clustering of innovations. Technologies form families, clusters, trajectories—and the main member of each family could be termed a basic innovation in the sense discussed earlier. The computer family requires, among many other items, microelectronics, memories, printers, software, visual display units; the motor car needs tyres, tarmac, petrol pumps, brakes, piston rings, etc.; the photographic process is associated with cameras, lenses, colour film, slide projectors, exposure meters, electronic shutters, etc. These are just a few examples of the interdependence of technologies and technological innovations.

A major problem of diffusion is the absence of infrastructure and of necessary members of the family in the early stages of a complex new technology. There is little advantage in having a telephone if nobody else has one; little advantage in having a robot if nobody can maintain it, and little advantage in having a car when petrol is unobtainable. On the other hand there is great disadvantage in not having a telephone when nearly everybody has one and great disadvantage in not using robots if all one's competitors use them. In the diffusion of a new technology the early stages are the hardest and success breeds success until, of course, obsolescence and increased competition make the innovator's life hard yet again. Success of the product may accelerate, but a growing number of manufacturers will chase this success and often murderous competition and market saturation will cause great difficulties and a misallocation of resources.

SUMMARY OF TENTATIVE CONCLUSIONS

To navigate between the Scylla of technical failure and the Charybdis of commercial disaster needs a helmsman with a keen eye for each. The odds are weighted heavily against success and only a product champion and a team who are both competent and determined can shorten the odds. When it comes to the frequent decisions on whether or not to continue with the development of the innovation, the problem often becomes one of determining the boundary between perseverance and obstinacy.

There is not, and cannot be, a universal prescription for success in innovation. Because of the vast range of markets and the practically infinite range of products and processes, no simple rules or procedures can be expected. The only universal, albeit unhelpful, recommendation is to choose horses for courses. All that innovation theory can achieve is to provide pointers, insights, signposts and food for thought. Particularly the realisation of the different phases of innovation and their characteristic requirements may yield at least some very general guidelines to be considered in each instance.

The zeroth phase, which needs to be borne in mind and reviewed at all times, simply requires a conscious appraisal of the capabilities of the firm and the dangers and opportunities offered to it by the outside world.

The requirements of the first phase can also be summed up in a few slogans of rather deceptive simplicity: keep your eyes open, keep good internal and external communications, keep a wide network of contacts and a stimulating open environment in the hope that you will spot a technical possibility–market need relationship before everybody else does. Clearly, much of the searching for such possibility-need constellations will occur around the current major areas of development—in the current mainstream trajectories of new technology. In these areas no doubt opportunities exist, but these are also the areas where everybody is searching. At the time of writing, the main avenues of innovation, as it were, are microelectronics/information technologies and bio-engineering. No doubt every firm needs to ask itself whether these new technologies are relevant to it, but need not worry if they are not.

The second phase is the hardest to sum up. This is where the skills of the management team are tested to the utmost. For they must bring together all the required resources into the right constellations at the right time. The team must be competent and dedicated, yet it must also be prepared to abandon the project or change direction radically. To be fully committed and yet detached, to marshall diverse technical resources and yet keep an eye on costs and eventual markets, to be secretive and competitive and yet know what is happening in the relevant part of the world— all these are necessary requirements and amount to a tough specification for would-be managers of innovation. In this phase too open communications and lack of disciplinary boundaries are self-evident prerequisites, for real problems do not fit artificial disciplinary boxes.

The third and fourth phases of an innovation come successively closer to normal management requirements and therefore subject to normal management techniques. What remains special is that every-thing—product, production method, organisation, skill require-ments, power structures, final market—remains fluid. As a result, the number of decisions to be made is very much greater than usual and total management effort needs to be appropriately large. It seems very helpful for the project management team to remain involved right into the fourth phase and then gradually to hand over to routine line management. At this point, the revolutionary phase is over and some semblance of normality should prevail again.

Perhaps Keynes's comments on industrial investment might form an apt postscript to our chapter on technological innovation, as it emphasises the spirit of adventure which alone enables people to embark upon such hazardous undertakings.[30]

In former times, when enterprises were mainly owned by those who under-took them or by their friends and associates, investment depended on a sufficient supply of individuals of sanguine temperament and constructive impulses who embarked on business as a way of life, not really relying on a precise calculation of prospective profit. . . . If human nature felt no temptation to take a chance, no satisfaction (profit apart) in constructing a factory, a railway, a mine or a farm, there might not be much investment merely as a result of cold calculation.

NOTES

1. The following works on the economics of technological innovation have been extensively consulted: Freeman, C., *The Economics of Industrial Innovation*, Harmondsworth, Penguin, 1974 (second edition, London, Frances Pinter, 1982); Jewkes, P., Sawers, D. and Stillerman, R., *The Sources of Invention*, London, Macmillan, 1968; Langrish, J., Gibbons, M., Evans, W. G. and Jevons, F. R., *Wealth from Knowledge*, London, Mac-millan, 1972; Mansfield, E., *The Economics of Technological Change*, New York, Norton, 1968; Nelson, R. and Winter, S. G., 'In search of useful theory of innovation', *Research Policy*, 6 (1977), pp. 36–76; Pavitt, K. (ed.), *Technical Innovation and British Economic Performance*, London, Macmillan, 1980; Parker, J. E. S., *The Economics of Innovation*, London, Longman, 1974; Rosenberg, N., *Perspectives on Technology*, Cambridge University Press, 1976.
2. Braun, E., Moseley, R. and Wilkinson, B., 'Manufacturing innovation in the West Midlands material forming industry', *Omega*, 9, No. 6 (1981), pp. 563–70.
3. Braun, E. and Macdonald, S., *Revolution in Miniature*, Cambridge University Press, second edition, 1982.

4. See Parker, *The Economics of Innovation.*
5. Braun and Macdonald, *Revolution in Miniature.*
6. See e.g. Samuelson, P. A., *Economics*, Tokyo, McGraw-Hill KogaKusha, tenth edition, 1976, p. 537.
7. See e.g. Rothwell, R. and Zegveld, W., *Technical Change and Employment*, London, Frances Pinter, 1981.
8. Green, K. and Morphet, C., *Research and Technology as Economic Activities*, London, Butterworths, SISCON series, 1977.
9. See Mensch, G., *Das technologische Patt*, Frankfurt, Fischer, 1977; and Freeman, C., Clark, J. and Soete, L., *Unemployment and Technical Innovation*, London, Frances Pinter, 1982.
10. Haustein, H. D. and Neuwirth, E., *Long Waves in World Industrial Production, Energy Consumption, Innovations, Inventions, and Patents and their Identification by Spectral Analysis*, Laxenburg, International Institute of Applied Systems Analysis, Working Paper 82-9, January 1982.
11. Freeman, C., Clark, J., and Soete, L., *Unemployment and Technical Innovation.*
12. Braun, E., 'Constellations for manufacturing innovation', *Omega, 9*, No. 3 (1981), pp. 247-53.
13. Freeman, *The Economics of Industrial Innovation*, p. 176.
14. Braun, E., Collingridge, D. and Hinton, K., *Assessment of Technological Decisions*, London, Butterworths, SISCON series 1978.
15. Ibid.
16. Braun, E., 'From transistor to microprocessor', in Forrester, T. (ed.), *The Microelectronics Revolution*, Oxford, Blackwell, 1980, pp. 72-82.
17. See Braun, Moseley and Wilkinson, 'Manufacturing innovation in the West Midlands'; and Wilkinson, B., *Technical Change and Work Organisation*, Ph.D. thesis, University of Aston, 1981 (published in abridged form as *Shopfloor Politics of New Technology*, London, Heinemann, 1983); and Sorge, A., Hartmann, G., Warner, M. and Nicolas, N., 'Microelectronics and manpower in manufacturing: applications of computer numerical control in Great Britain and West Germany', Berlin International Institute of Management discussion paper, October 1981.
18. Cantley, M. and Sahal, S., *Who Learns What*, Laxenburg, International Institute of Applied Systems Analysis, Working Paper 79-110, 1979, and Jackson, D., *Introduction to Economics*, London, Macmillan, 1982, pp. 337-62.
19. See Langrish, Gibbons, Evans and Jevons, *Wealth from Knowledge.* One of the better known empirical studies of innovation was project SAPPHO, described in Freeman, *The Economics of Industrial Innovation*, pp. 171-97; see also 'Report on Project SAPPHO' (1971) and 'Success and failure in industrial innovation' (1972), Science Policy Research Unit, University of Sussex.
20. Freeman, *The Economics of Industrial Innovation.*
21. Fleck, J., The Adoption of Robots, 13th International Symposium on Industrial Robots, Chicago, April 1983; Fleck, J., *The Introduction of Industrial Robots*, London, Frances Pinter, forthcoming; Scarbrough, H.,

The Control of Technological Change in the Motor Industry, Ph.D. Thesis, University of Aston, 1982.

22. Braun and Macdonald, *Revolution in Miniature*, pp. 109–10.
23. Elsom, S., 'Protecting software against piracy', *Data Processing*, 2, No. 3 (April 1983), pp. 6–8.
24. Zermeno, R., Moseley, R. and Braun, E., 'The industrial use of robots', *The Industrial Robot*, 6 (1979), pp. 141–7.
25. Freeman, *The Economics of Industrial Innovation*, p. 177.
26. Braun, E., 'The science–technology interaction', in Williams, B. (ed.), *Proceedings of Conference on Technical Change* held at Technical Change Centre, London, February 1982, forthcoming.
27. Ravetz, J. R., *Scientific Knowledge and its Social Problems*, Oxford, Oxford University Press, 1971.
28. Braun and Macdonald, *Revolution in Miniature*, pp. 186–204.
29. Ibid., pp. 202–3.
30. Keynes, J. M., *The General Theory of Employment, Interest and Money*, London, Macmillan, 1967, p. 150.

3 Contemporary Fears about Technology

Fear of technology is not new. Prometheus was punished for stealing fire with rightfully belonged to the gods; Icarus paid with his life for presumptuously reaching for the sky. The ancients feared punishment for looking beyond their natural domain; early industrial man feared the machine for encroaching upon his own domain of skilful employment. The machine threatened jobs and took over human skills, while the factory became the galley to which man, woman and child were slaves. Some contemporary writing of eighteenth- and nineteenth-century England is uncannily like modern writing on the topic of de-skilling and the threat of unemployment. For then as now, machines could perform sequences of motions without human intervention. Then as now, machines and the division and organisation of labour associated with the factory system greatly enhanced the product per worker. The result is the redundancy of some skills previously held in high esteem and a decreased demand for labour, unless total demand for goods and services rises to meet the new productive potential. At a meeting of the Co-operative Congress on 23 April 1832

Mr. Pare (from Oldbury, near Birmingham) rose to second the resolution. The present distress among the industrious classes may be attributed, in a great measure, to the vast increase of machinery, in all the manufacturing towns and villages, during the last fifty years. In the year 1792 the machinery then in operation in the country was equal to the labour of about ten millions of men; but now, mark the difference, and your wonder, at our present distress and want of employment, will soon cease. At the present time, the machinery— and I have made my calculations from good authority—in the United Kingdom alone, is equal to the laborious exertions of six hundred millions of hard-working industrious men.[1]

Opposition to machines came mainly from spinners and weavers, who felt directly threatened, and worker politicians, who felt a general opposition to the new type of oppression, misery and exploitation to which the factory worker was subjected. A secondary source of opposition was the romantic defender of the Arcadian present against a barbaric future. Vague fears about the machine-ridden future were expressed even by the great Goethe himself, 'Das über-hand nehmende Maschinenwesen quält und ängstigt mich' ('The growing prevalence of machinery worries and frightens me').[2]

The opposition to machines never reached a strength which might have seriously threatened their progress. Sufficient numbers of people benefited directly or indirectly from the new system of manufacture to give it powerful backers. And even the new proletariat presumably found life in factories at least marginally more palatable than life as a pauper. Gradually it became apparent that machines increased total wealth and political opposition changed from hostility towards machinery to hostility towards its owners. Socialism, from its early beginnings, opposed the owners of the means of production but never the means themselves. Indeed the creation of a strong foundation of wealth, based on efficient manufacture, was—and remains—a cornerstone of socialist faith. The controversy was about distribution of wealth and control of the means of production, not about the nature of technology.[3]

During the 1930s a wave of fear about the use of science and technology emanated from imaginative writers whose main concern was the loss of personal freedom. At a time of rising dictatorships, which made full use of available technology, the preoccupation with visions of a horrific future of people enslaved by technocratically efficient masters was highly plausible. Names such as Čapek, Huxley and Orwell spring to mind with a vision of science and technology getting out of control and being unscrupulously misused to destroy all that is essentially human. Efficiency triumphs and human values fall before it. The dreadful period culminated in the horrors of war, of extermination camps and finally the atom bomb.

Yet despite this, many of the fears of science were forgotten during the years following the Second World War. The general mood was one of a strong belief in the power of science and technology to create a better world. Not only would the material circumstances improve and poverty be banished, but science and the scientific way of thinking would lay the foundations of a better ordered rational world of international cooperation.

Peaceful uses of nuclear energy, vast new research institutes for everything from elementary particle physics to molecular biology, a huge expansion of higher education with special emphasis on science, all these were symptoms and symbols of renewed belief in the power of science and technology. By that time the two concepts, starting from quite different origins, had become very closely linked and were often used interchangably, with science as an umbrella term embracing modern technology.

The mood of optimism soon gave way to one of horror of modern science-based weapons in a re-arming world, yet this horror was coupled with continued faith in the peaceful utility of science and technology. A bifurcation occurred—faith in the peaceful use of science and horror of scientific warfare. But even the faith in peaceful uses of science and technology did not remain unchallenged for long. By the early seventies there was a veritable explosion of writings expressing fear for the future well-being, even the future existence, of mankind unless the development of science and technology were bridled.

Quite apart from the overriding fear of mankind destroying itself in nuclear war—a fear so awesome that most people shrink away even from the fear and prefer not to contemplate the gruesome prospect—a whole range of worries about technological developments began to raise a powerful voice.

The main strands of the argument in the early seventies are easily summed up:

The combined use of improved health care, especially for infants, and improved agricultural methods led to a population explosion. While science could provide means of contraception, these were not readily taken up in the poorer parts of the world and thus death rates had decreased much faster than birth rates. The world was being overrun by people and, in the long term, some kind of Malthusian disaster was bound to occur. Although fear of the population explosion was mainly directed at the developing countries and only indirectly related to science and technology, it nevertheless highlighted the inability of scientific methods to solve the problem of overpopulation, and the inadequacy of science-based agriculture to ban the spectre of hunger.

The fear of population growth overtaking the earth's capacity to feed it, and the fear of running out of adequate resources for a growing population, is an old preoccupation of economists. Malthus and Ricardo were primarily concerned with limitations imposed by the then only source not only of food but of wealth—the land. Later economists, such as Jevons in 1865, foresaw similar limitations imposed upon the creation of industrial wealth by insufficient coal reserves.[4]

The second main fear was environmental pollution. Although gross contamination of the environment is a problem as old as concentrated human dwellings, in the past it was caused by lack of technology. Night soil was conquered by sewers, horse-dung by the motor car. Modern pollution, on the other hand, is caused by excessive or incautious use of technology and is of a more frightening and less readily understood nature. There is something natural and almost reassuring about excrement, while radioactive or chemical wastes are mysterious and sinister.

The long-term use of even beneficent fertilisers and insecticides was proving damaging to wildlife and to rivers and lakes; the effluent from households and factories not only disfigured but often positively poisoned the waterways; the emissions from factory and domestic chimneys and, worse still, millions of car exhausts threatened to poison the very air we breathed.

Fears of an even more ominous kind were voiced, fears about stable gases, such as carbon dioxide, being released into the atmosphere and causing permanent climatic changes which would eventually put an end to life on earth. The ecosphere—the subtly balanced web of organisms and their life support systems—might become damaged beyond redemption.

The third fear was that of running out of natural resources. Despite all the difficulties inherent in defining what a natural resource is and even greater difficulties in knowing the true reserves of whatever resource, it is obvious that resources are finite and that ever increasing consumption will eventually exhaust at least some of them. And despite the numerous times such fears have proved unfounded since the time of Malthus, their simple logic compels attention time and again. The greatest fear concerns, of course, reserves of energy, especially oil. Energy occupies a special position because it is central to all technical activities and, while the availability of raw materials can be stretched by recycling, there is no way in which energy can be re-used once it has been consumed.

The numerous writings of the time bear witness not only to the concerns and fears, but also to the impossibility of resolving the fierce controversies by scientific means.[5] Even recourse to the computer did not resolve the issues. While a computer model designed by competent scientists showed convincingly what could have been shown by simple argument, that exponentially increasing consumption of finite resources and exponentially increasing loading of natural waste disposal capacity would eventually lead to some form of decline or collapse;[6] this model and its assumptions were fiercely criticised by other equally well-qualified scientists.[7] The debate moved firmly in the realm of what Weinberg has called trans-science, a realm in which beliefs and values colour the debate and undisputed facts are inadequate to give decisive consensus victory to one side or the other.[8]

Not unnaturally, the scientific community reacted with frantic efforts to establish more facts. Scientists tend to believe that controversy is normally based on ignorance and that, provided sufficient scientific research of the right quality is carried out, the true facts will emerge and put paid to dispute.

Large-scale research programmes and a flurry of smaller-scale projects set out to establish the true dangers of various pollutants to human, animal and plant life, to assess the risks to the ecosphere, and to ascertain the availability of reserves of natural resources. At the same time, numerous attempts were started to develop better computer models, which were to relate the various facts to each other and improve our understanding of the multiple dependencies among human activities and their impact upon the environment.

About a decade later, some of these questions had been resolved, most had proved to be long-term research projects, and the fierce debates of the early seventies had given way to equally fierce debates, albeit mostly on different topics.

The inability of science to resolve conflicts relating to scientific and engineering matters has caused much heart-searching. For is not science 'public knowledge', consensus by all those informed enough to understand?[9] This may be the case—although there are vitally important exceptions even to this concept—but the strength of science lies in solving self-selected 'puzzles'.[10] Not only are some questions too difficult to resolve with reaonable effort in a reasonable time, but there is a large range of questions unsuited to scientific resolution even if they appear to be of a scientific nature.

As an example we may take the question of whether lead should or should not be used as an additive to petrol. Some of the facts in this case are simple and undisputed. Lead, ingested in large quantities, is a highly toxic substance causing definite and well-documented clinical symptoms. Lead additives to petrol do improve the burning qualities of the fuel (the so-called octane rating) and thereby increase the efficiency of engines. Against these two facts, a very large number of open questions has to be faced. What, if any, is the lower limit of lead ingestion at which it becomes harmless to human health? What fraction of total lead intake by humans must be ascribed to car exhausts and what to other sources? How is the lead distributed over the population? Do some groups, such as urban children, face special risks? These are some of the questions science may eventually answer and has already partially answered. But what of the equally difficult questions in which technology and economics are hopelessly jumbled? Is the addition of lead to petrol the best, or even the only, means of making car engines more efficient? Do we reduce total petrol consumption and total costs by using these additives? Can existing oil refining facilities cope with the production of fuel without lead additives? Why cannot all motor manufacturers produce engines which do not require high octane fuel and are such engines more wasteful? Would not a general speed restriction yield much better economies than all the lead additives put together?[11]

The last question shows how close the complex of unresolved issues comes to politics. For not only do economic, scientific and technical questions require unequivocal answers, but any solution to the lead problem must be politically acceptable. A severe speed restriction or limit on engine power apparently is not. In recent years, some compromises between motor manufacturers, oil companies and governments have been achieved and lead levels are slowly coming down, despite the fact that the true benefit of the reduction is not scientifically established. In view of the long-term nature of the scientific enquiries and of the possible hazards caused by lead, there really is no option other than a decision in the face of ignorance. By the very nature of such a decision, it is likely to be one of a more or less prudent compromise between short-term economic interests and possible long-term health hazards.[12]

Environmental pollution by agricultural chemicals shows similar features: a small number of chemical manufacturers producing the materials; a large number of farmers individually deciding to use them in their own economic interest; damage caused to waterways by chemicals being washed into them; and a government trying to balance the interests of industry, farmers, and a population that requires not only cheap and plentiful food but also clean and healthy waterways. Again a large number of unanswered scientific and economic questions arise. Is it possible and economically feasible to clean up the waterways with undiminished use of fertilisers and pesticides? Is it possible to develop alternative and less harmful agricultural chemicals? Should the quantities and types of chemicals be restricted? What about damage other than to waterways, such as direct damage to wildlife? Is it possible to harm only harmful insects and leave useful ones unscathed or will any interference in the balance of nature be harmful, or at least problematic? One of the political solutions already attempted has been the banning of DDT everywhere except in regions of acute danger from malaria and other insect-borne diseases.[13]

When it comes to questions of global damage to the ecosphere, our ignorance knows no bounds. It has been variously argued that as increasing amounts of fossil fuels are burned, so the concentration of carbon dioxide in the atmosphere will increase and this will result in an effect similar to the heating of a glasshouse. The short wavelengths radiated from the sun will still penetrate the atmosphere, but a higher proportion of the longer wavelengths radiated by the earth will be absorbed in the atmosphere by the higher concentration of carbon dioxide. Thus the balance of radiation will change and the average temperature of the earth will rise, with possibly catastrophic consequences, such as the melting of the polar ice caps and flooding of large parts of the inhabited earth. Years of scientific effort aided by sophisticated computers, have not as yet provided an answer as to the reality of this supposed danger.[14]

The question of the exhaustion of natural resources runs into yet another set of difficulties. For one thing, the very definition of a natural resource is problematic. While flint was a major resource of stone-age man, it is hardly very important now. On the other hand, silicon has only very recently achieved the status of a resource. With improved techniques of extraction and purification, it is possible to use ever poorer grades of metallic ores. In principle, this requires ever greater inputs of energy, but again improved techniques can, albeit only partially, overcome this problem. While domestic production of copper in the United States used ores of average copper content of between 1.2 and 2 per cent up to 1940, the average content came down to about 0.6 per cent by 1965, with no increase in the cost of production. The energy crisis of 1973 has, of course, drastically altered the situation and production costs have become more sensitive to energy requirements. The share of the primary metal-producing industries in total energy consumption is substantial. In the United Kingdom in 1968 the production of iron and steel, excluding mining, accounted for 12.5 per cent of total energy consumption, while non-ferrous metals used 1.2 per cent. For comparison, road transport accounted for 14.1 per cent and domestic use for 26.3 per cent.[15]

If there are problems of definition in the determination of what constitutes a raw material or a suitable ore, there are even greater problems in defining known reserves. There is little incentive to prospect for materials which will not be used within twenty years or so. Consequently, there is a tendency for reserves to be sufficient for twenty years at all times.[16] Prospecting is largely carried out by private concerns and it is very hard even for governments, let alone for non-governmental researchers, to know the true state of reserves. In countries where mineral extraction is a state monopoly, the state is as reluctant as its private counterparts elsewhere to disclose what facts are known.

There are two ways in which reserves can be further extended: recycling and substitution. In the case of copper, for example, well over 20 per cent of the material used in Britain comes from scrap metal. The proportion can be increased, but severe problems of re-use arise when the material is widely dispersed in small quantities or contaminated by other materials. Substitution of more common materials for scarcer ones is another way of extending the life of a rare resource. An important example is the substitution of ubiquitous aluminium for rare copper. Clearly, each material has some unique properties and there are technical and economic costs to substitution, but some scope exists.[17] *In extremis*, such as in war-time Germany, a great deal of otherwise uneconomic substitution was undertaken.[18]

Although in principle it is undoubtedly true that exponential growth of consumption is irreconcilable with finite resources, the practical implications of this truth are far from clear. Some materials are available in such high average concentration that any rock containing a little above average becomes an ore. Other materials, notably gold and mercury, need enormously enriched rock to be useful as an ore. Some enrichment factors are shown in Table 3.1, but these are determined as much by the technology of extraction and the price of the material as by nature. No doubt those materials in highest demand which are available in the smallest total amounts in the earth's crust are likely to become extremely scarce soonest.

The most central fear about exhaustion of natural resources is the fear of running out of energy. While other materials may be recycled or substituted for, or goods made to last longer thus reducing

Table 3.1. Enrichment factors for some common metals

Metal	Per cent in crust	Per cent in ore	Enrichment factor
Mercury	0.000008	0.2	25000
Gold	0.0000002	0.0008	4000
Lead	0.0013	5.0	3840
Silver	0.00007	0.01	1450
Nickel	0.008	1.0	125
Copper	0.006	0.6	100
Iron	5.2	30.0	6
Aluminium	8.2	38.0	4

Source: National Research Council, *The Earth and Human Affairs*, San Francisco, Canfield Press, 1972, p. 80.

the need for new materials, energy resources cannot be stretched in quite the same way. It is true that there is a law of conservation of energy, which says that energy can only be converted from one form into other forms but can never be lost. The law is as true as it is unhelpful, because energy, though conserved, is degraded in use into dispersed heat and in this form cannot be re-used. The friction between motor car and road, which is overcome by the engine deriving energy from fuel, heats up both tyres and road and therefore the energy remains in existence, but it is impossible to regain this energy and re-use it to drive the car a little further.[19]

Energy is central to all industrial activities and, in most climates, to human survival. In fact manufacturing activity can be defined as the use of energy to convert raw materials into useful objects. Raw materials may vary and the useful objects may differ, but energy is always at the heart of the manufacturing process. The use of energy from sources other than human or animal is an essential feature of industrial society. Energy from mineral sources enabled manufacture to grow far beyond the limits imposed by animal power; it enables fast mass transportation of goods and people; and it makes urban settlement possible.

It has been argued that there exists a linear relationship between economic growth and the consumption of energy. This argument has been successfully refuted and it is now accepted that economic growth is possible with a lower ratio between gross national product (GNP) and total energy consumption than was previously regarded as necessary. However, measures to save and conserve energy have limited scope and eventually some relation between energy consumption and economic growth will reassert itself. It seems likely that the scope is very considerable, for past use of energy can only be described as profligate, but all the same, adequate energy supplies are vital for the survival of technological society.[20]

The true state of reserves is hard to assess and the twenty-year rule applies to oil more than to any other commodity. Nevertheless, most people believe that oil reserves are finite in a very real sense and that oil will become scarce in the foreseeable future. No doubt more oil will be discovered and the reserves can be stretched by several methods: more efficient extraction, leaving less oil in the ground; the use of oil shales and tar sands, of which there are considerable reserves (but their oil content is low and costly to extract); conversion of other fuels, especially coal, into oil. But when all is said and done, in the long run alternatives to oil will have to be developed and the talk is of alcohol, hydrogen and electric batteries. Which, if any, of these alternatives will eventually dominate is anybody's guess. It seems probable that all the alternatives, from oil to electricity, will be used as mobile fuels, each capturing a niche of the market appropriate to its special advantages. Fuel will probably never be as cheap again as it was before 1973; in fact, despite the present halt in the rise of fuel prices, it is likely that rising costs will force an end to profligacy.

Energy requirements for purposes other than road, air and sea transport pose somewhat different problems. Some countries, notably the USA but also Great Britain, still have substantial known reserves of coal and these will supply a good proportion of electricity generating requirements for well over a hundred years. Coal may also compete with oil as a raw material for a whole range of chemical products, particularly plastics. In 1974, 4.9 per cent of all industrial energy consumption in the USA was accounted for by the use of oil as feed stock for chemical production.[21]

As we approach the question of energy supplies more closely, the impression of opening Pandora's box gains strength. The controversies are fierce, complex and emotion ridden. The very fact that energy is at the core of technological society makes the debate about the future of energy supplies a central, hotly and widely debated issue. In the present context, we can do no more than provide a very brief summary of the debate.[22] The key issues are:

(i) it is impossible to forecast energy requirements accurately into the future;
(ii) the lead times of building new electricity generating plant, coal mines, hydroelectric stations or any other major energy supply installations are very long;
(iii) because of the central role of energy, and the problems of difficult forecasting and long lead times, planners generally prefer to err on the side of caution. The result is an over-capacity of supply, which tends to encourage excessive use in an attempt to utilise the capacity provided;
(iv) there is a strong belief among many people that renewable sources of energy can be developed, such as wind power, wave power and, above all, solar power.[23] The same people tend to believe also that energy conservation can reduce total demand very considerably;
(v) most governments and central authorities in charge of energy supplies, backed by scientists and engineers associated with either traditional sources of energy or with nuclear power, do not strongly share these beliefs. They believe that renewable sources of energy will play only a limited role and energy

conservation will reduce further requirements only marginally. They conclude that the only viable and secure long-term solution to the energy problem is nuclear energy in one form or another;

(vi) nuclear energy arouses the fiercest controversies of all. Its protagonists hail it as the only realistic solution to the energy problem and see in it a safe and reasonably clean long-term source of energy. They point out that radioactive materials can be kept under control, whereas the emissions from fossil-burning power stations cannot. The opponents of nuclear energy see in this technology a severe threat to mankind because of the possibility of leakage of radioactive materials, the danger of misuse of fissionable uranium for non-peaceful purposes, and the elaborate security arrangements required to safeguard radioactive materials. They fear also that major malfunctions with catastrophic explosions and radioactive leaks might occur.[24]

So strong has the opposition to nuclear energy become in some countries that it transcends the fear of this particular technology and embodies fear of modern technology in general. The sinister incomprehensible nature of the dangers, the massive centrally planned and controlled plants, the lack of local personal involvement in any decisions—all these may be factors in the opposition. Even more deeply rooted antagonism towards what some see as sterile materialism and a lack of a human face in modern science-based technology may find expression in opposition to nuclear energy. A strong belief that no generation should burden its successors with a heritage of radioactive waste products, which remain dangerous in perpetuity, causes many people to join the ranks of opposition to nuclear powers. A vague association with the horrors of nuclear warfare no doubt further reinforces the lobby against nuclear energy and turns it into overall opposition to all things nuclear.

The lack of a human face to technology is ever more strongly felt. The expression is probably no more imprecise than the worry. People working in factories often perform extremely tedious and monotonous tasks and are being paced by a machine instead of pacing themselves or even being driven by a human boss. The relentless, endlessly repetitive nature of machine pacing as against the somewhat erratic human pace is probably one aspect of the feeling. Another is the removal of human contact. People on production lines often cannot talk with each other and machines have become so powerful, efficient and automated that humans no longer need to cooperate with each other to achieve the set task. The relentless drive for efficient production, with its emphasis on division of labour and control over all the work process, has removed all human give-and-take from routine production. Add to this the anonymity of the big city, with people in motor cars shut off from their fellow humans and people watching television not communicating with anybody. When then even information clerks are replaced by television screens or synthesized voices mindlessly repeating the same message, whether comprehensible or not, the inhuman nature of technology becomes a little obtrusive.

Though the debates have shifted a little over the years and nuclear power rather than the natural environment has taken the limelight, the dangers to the latter have, sadly, not passed. Indeed acid rain is now a topic of major concern and the great forests of Europe must be regarded as an endangered species. Sulphur and possibly nitride emissions are regarded as the culprit, but as the emissions do not recognise the maze of national frontiers of Europe, their control has become a matter of prolonged political wrangling. Whether death will overtake the forests while politicians are talking is a matter of conjecture. The clock of destruction is certainly ticking away. The destruction of forests by technology is especially painfully ironic, for the proper management of forests was one of the first lessons man learned from the results of his previous wanton exploitation.

More recently, even civil engineering works, which were long regarded as a cause of pure pride, have become objects of worry and protest. They are now often seen as destroyers of nature and a symbol not so much of man's conquest as of destruction of his natural habitat. Power stations, bridges, dams and motorways are no longer regarded as monuments to human ingenuity and landmarks in the successful march of civilisation. On the contrary, many now view these works as monuments to Man's folly and his despoiling and abuse of nature.

Man alone requires little space and makes little impact upon the natural environment. But Man armed with technology requires a great deal of space—his radius of action has multiplied manyfold—and covers vast tracts of land with stone, brick and concrete. As the population increases and vast masses of people mill around their concrete jungles, filling every available space with their motor cars, an immense fear rises, an impression that Man is overrunning the earth and turning its lush beauty into a barren hostile desert. Hence the protests against every new motorway and new runway or new power station. It may be true that the protesters often drive hundreds of miles in fast cars to their marches, but the basic fear and unease is there despite all inconsistencies and follies. There is much scope for enlightened policies and politics.

SOCIAL IMPACT OF MICROELECTRONICS

The fear of the implications of microelectronics is the most recent and most topical fear of technology.[25] It has three major components:

(i) fear of unemployment;
(ii) fear of loss of skills;
(iii) fear of loss of privacy and freedom.

(i) Fear of unemployment

Not unreasonably, the fiercest debates about the future of energy raged after the oil crisis, while the worst fears about technological unemployment are being voiced now, at a time of economic crisis and high unemployment. The argument in its simplest form states that as microelectronics takes over more and more control functions in manufacture, and information handling functions in administration, so human labour will become increasingly redundant. Clearly, if machinery becomes ever more efficient, the same amount of goods and services can be produced with a smaller input of labour.

At the next stage of sophistication the argument goes on to acknowledge that there had been a very great decrease in the proportion of labour employed in agriculture ever since the industrial revolution. The surplus labour was taken up by mining, transport, manufacture and services. In recent years, when agricultural employment had reached a minimum level, and employment in mining and manufacture also began to decline, the surplus was taken up by service employment (see Table 3.2). However, if new technology greatly increases efficiency in both manufacture and services, then there is no sector left which could absorb the surplus labour. The spectre of fully automated factories producing goods untouched by human hands and of offices with vastly efficient computers and word processors replacing armies of clerks and typists, while unemployment becomes the norm and work the exception, haunts many a writer and speaker on the impact of microelectronics.

The strength of the above argument lies in its recognition that technological change brings with it inevitable shifts in patterns of employment. This is true both between major sectors of the economy and within each sector. So for example the change from town gas, produced in a coking process from coal, to natural gas found

Table 3.2. Employment structure in Germany, France and the U.K., 1963–1978

Proportions of total employment		1963	1968	1973	1978
1. Primary	G	4.3	3.2	3.0	2.7
	F	8.0	5.8	4.0	3.0
	UK	6.1	4.4	3.5	3.2
2. Manufacturing	G	44.2	43.8	40.8	39.4
	F	35.5	32.9	33.3	30.8
	UK	37.0	36.4	34.3	32.2
3. Construction/utilities	G	10.7	9.8	10.3	8.7
	F	11.1	12.0	11.2	9.9
	UK	8.2	8.6	7.6	7.2
4. Distribution, hotels, restaurants	G	12.6	13.1	13.0	12.7
	F	12.2	13.3	13.7	14.4
	UK	15.8	15.2	15.6	16.0
5. Producer services	G	10.3	11.0	11.6	11.8
	F	11.6	12.6	13.3	14.7
	UK	11.4	11.6	12.2	12.4
6. Other personal and collective services	G	18.0	19.1	21.3	24.6
	F	21.6	23.6	24.2	27.2
	UK	21.1	23.8	26.3	29.2

Source: After Gershuny, J. I. and Miles, I. D., The New Service Economy, London, Frances Pinter, 1963, Table 2.2.

in the North Sea, has reduced the number of manual workers employed by British Gas from 112,121 to 40,674 in the years 1952 to 1976. During the same period, British Gas increased their staff by 25,142 and, no doubt, contractors employed to adapt equipment to the new fuel also employed a good number of people.[26]

The main flaw in the argument is the lack of recognition of the fact that human needs, or at least many human needs, are insatiable. Here lies one of the differences between agricultural production and industrial production. On the supply side, agricultural production was at one time thought to be severely constrained by availability of land. It turned out that land could become much more productive by the addition of industrial products, such as chemical fertilisers and tractors, so that the supply of food in advanced countries defied Malthus, especially as the growth of population limited itself for reasons other than hunger. On the demand side, only the quality and commercial value of food consumed could be increased, often by adding industrial processes, but total consumption remained severely limited by the capacity of the human body. In economic terms, the marginal utility of purchasing food beyond one's own capacity to eat it rapidly approaches zero.

Not so with industrial products and with services. The person who already owns a washing machine may not want another, although in due course he or she may be persuaded to exchange an obsolescent model for the latest design. Similarly, the person who owns one car may not want another, although a family may be persuaded to have two or more cars and obsolescence again takes a heavy toll. However many pairs of shoes, or records, somebody may have, he or she may easily be induced to buy more. And however many gadgets a person has, there is always something new, something better, or something he or she has not got yet. The very rich may have to go as far as buying an echo sounder for their yacht or a crystal decanter for their picnic hamper, while the less affluent will have to be content with a ball-point pen equipped with a digital watch.

Demands for services are even more insatiable. Having been to a nice restaurant yesterday prevents no one from going today, provided they can afford it. The same applies to opera, the theatre, football matches or holidays, all depending on taste, financial state and other circumstances. Apart from services aimed at private enjoyment, there is also a large range of public or private welfare, educational and utility services which are pretty well insatiable. Among them are health, care for the aged or underprivileged, education and training at all levels, including continuing education, public transport, sport and recreation facilities, and the maintenance and renewal of urban amenities.

In a sense leisure time is also a consumable item and most people can happily use more than they have, provided that they have enough money, that there are enough facilities, and that leisure is a change from work and not a substitute for it.

Looking around the world, it certainly does not appear as if we were in any danger of running out of needs to be satisfied or work to be done. Indeed, to the millions of underprivileged people in advanced countries, not to speak of developing countries, all that has been said so far must sound like a bitter caricature of reality. Although changes in technology demand changes in the jobs people do and may cause considerable dislocation and disturbance, it seems unlikely that technology itself should determine levels of employment or unemployment. No doubt technology is a factor, but only one of many factors which determine economic activity.

The level of employment in an economy is dependent upon innumerable factors, such as the state of the world economy, aggregate demand, inflation, educational capacity and labour relations, as well as on the technologies used and the rate of technological change. To elevate technology above all else and make employment forecasts on the basis of this single factor seems the height of, albeit fashionable, folly.[27]

Employment is not determined by the sum total of economic needs, but by the sum total of economic activity aimed at satisfying some of the needs. Total activity varies in irregular cycles and the causes of economic upswings and downturns are the subject of a vast literature in economics.[28] Various cyclical features of the economy have been identified, among them cyclic features of technology. In order to escape the unemployment implications of economic downturns, as well as for several other reasons, there is a relentless pressure for technological innovations and this leads us to one of the basic dilemmas about technology, the dilemma of rapid technological change. To provide economic growth and to satisfy an ever widening range of needs, technological innovation must proceed as fast as practicable. Yet to conserve natural resources and to avoid dislocations in the labour market, technology should evolve at a much steadier pace. What this pace should be and how to control it is a central problem of technology policy.

Total employment in an economy depends on the quantity of labour on offer, the amount of goods and services consumed, and the productivity and equilibrium of the economy. In principle, higher productivity enables more needs to be satisfied, which means more goods and services will be offered and the necessary means of exchange—money—will be available. The principle is unfortunately too simple to apply in reality, as all kinds of imbalances and malfunctions arise in an economy. Imbalances in external trade, imbalances between supply and demand because of structural weaknesses, property and financial speculation which diverts investment and causes imbalances, and inflationary pressures owing to a variety of causes. As no economy is closed, external economic ills spread like infections and political crises have profound effects upon economic events.

For all these reasons the correct analysis of an economic situation is fraught with difficulties and it is rare indeed for economists to reach unanimous conclusions. It is therefore not surprising that the debate on the employment effect of microelectronics has brought forth extremes of opinion—from the view that the impact would be catastrophic to the view that it would be negligible. It must be added that the extremely pessimistic views do not generally emanate from those with any profound knowledge of economics—expert economists tend to hold more moderately pessimistic or downright optimistic views in this matter and do not tend to overrate the role of technology as a cause of employment.

Much of the argument about microelectronics and employment tends to centre on the question of productivity. On the whole, microelectronics tends to increase labour productivity in both manufacture and services, while at the same time not opening up large new areas of economic activity or providing a large new range of consumer products. The production of electronics and the goods associated with it tends not to be very labour intensive. As a result, the direct labour displacing effect is expected to be greater than the direct work creating effect. In order to utilise to the full the productivity enhancement caused by the new electronics, it will be necessary to obtain structural changes which will make it possible to satisfy hitherto unsatisfied economic needs. Hence the universal cry for product innovations and also a fairly widespread belief that services must be enhanced and more leisure time must be given. Unfortunately some of the services that are known to be required, including housing for the lower income groups, public transport, health and education services, are difficult to finance privately and many countries have run into considerable resistance to increases in publicly financed services. In any case, structural changes are not easily achieved as there are many rigidities in an economic system and the achievement of a radically new equilibrium is a far from trivial requirement.

Despite misgivings about the possible effect of microelectronics upon employment, all industrialised countries feel compelled to introduce the new technology as fast as possible because of fears of falling behind in international trading competition. Modern goods, to compete on international markets, must incorporate the latest techniques, as otherwise they will not be wanted, and must be manufactured by the latest methods, as otherwise they will be too expensive and, perhaps, of inadequate quality.

Although it is very unlikely that the present employment problems, unevenly distributed throughout the industrial countries, have been caused by microelectronics or computerisation, it seems equally unlikely that microelectronics will provide the necessary impetus to move the economies into greater activity with higher employment. On the contrary, it is possible that growth in employment will lag behind general economic growth, if and when this resumes. The very large number of official and semi-official studies on the topic of the employment impacts of microelectronics all seem fairly unanimous in this view.[29]

Remedies must be sought in stimulating economic activity, including the renewal of decayed cities and the stimulation of all kinds of services, as well as industrial activity. If this is achieved, then the productivity gains and product improvements obtained by microelectronics will prove beneficial. If governments fail to achieve economic growth, then the productivity improvements cannot be absorbed other than in increased leisure time, whether properly distributed or meted out as unemployment to a politically poorly represented minority.

(ii) Fear of loss of skills

The ferocity of the debate on the influence of microelectronics on skills is second only to that on total employment. The essence of the argument is that because microelectronic equipment can perform more complex logic functions than mechanical equipment, so that complex sequences of events including 'decision' points can be programmed into machinery, machines will now incorporate skills previously vested in people. The apparent ability of computer controlled machines to decide in the light of circumstances which path to take, gives the machines powers which are uncannily like human brain power. In reality the computer can only do what humans have programmed it to do and becomes helpless if unforeseen circumstances arise, but within its own scope, the computer is incredibly fast and well-nigh infallible. Computer controlled machines can, like their mechanical predecessors, take over some functions which were previously within the human prerogative. The operator of the computer numerically controlled (CNC) machine tool does not require the same skills as the operator of a lathe or milling machine or any other hand controlled machine tool. This is only stating the obvious, as technology is the tool used by humans to satisfy their material needs and the nature of the tool clearly determines the skills required to handle it. If technology were a coin, utility and skill would be its two sides. There is no question that changes in technology cause changes in skill requirements. The only question is whether the changes are detrimental or otherwise.

If it were possible to measure the totality of skills employed in an economy, we would have to ask whether microelectronics will decrease the total or change its distribution or leave both total and distribution unchanged. Putting the question this way shows clearly that the answer can at best be conjectural. In mass production processes it is unlikely that the introduction of computer controls will decrease skill levels. Workers on production lines have long been deprived of skills and the short operational cycles and repetitive nature of the work take all meaning out of tasks which in themselves may require a certain knack. True skill must be associated with reasonably long work cycles and with a degree of discretion. It seems likely that in mass production processes computer controlled machines take over more of the unskilled tasks and leave to humans the supervisory jobs, rectification of machine errors, maintenance, repairs, programming. The real fear in mass production is that of massive reductions in the number of jobs, not in the reduction of skilled employment.

Batch production is quite a different story, particularly in small batches where the division of labour has not made individual tasks quite meaningless yet. As the programmable NC machine tool and the robot take over the production of small batches of products, so the numbers of skilled machine tool operators will decrease. The skills formerly embodied in these people may become incorporated in the machine programmes. A great controversy surrounds these questions. For the NC machine tool may be operated by a skilled person who programmes it, maintains it and supervises it or, alternatively, the operator may be replaced by a mindless machine minder and the programming and maintenance functions may be removed from the shop floor and handed to white-collar staff. In the former use, the skills of the machinists will be altered but not diminished; in the latter case the skills will be polarised; very little for the man on the shop floor, a great deal for the programmer and maintenance person.

This is perhaps the clearest example showing that the question of skills is a deeply political one. All groups within the factory system attempt to obtain as much control over the work process as they can and the acquisition and retention of skills is closely associated with control.[30]

To speak of the impact of microelectronics upon skill requirements is missing the point. The point is that every change in technology requires changes in skills and therefore opens a new phase in the struggle for distribution of skills and power. In the case of microelectronics, the main fear is that skills will become more polarised—in the extreme there would be only entirely unskilled workers and very highly skilled ones, with nothing in between— an army of privates and commissioned officers only, with no ranks to rise through. Many skills will become more abstract and less manipulative and there is a real possibility that many people with great manual dexterity and creativity will be deprived of the use of their gifts. To imagine, however, that currently production workers have a great deal of creative scope is indulging in fantasy. Micro-electronics will probably change things at the margin only, though probably for the worse, unless political action is taken.

The drift of labour away from direct production to service occupations, within and without manufacturing industry, has been proceeding for some considerable time, caused by increasing productivity of mechanised and automated machinery and increasing complexity of organisations and products. Technology may be said to have caused a structural change in employment from manufacture into services, including services within manufacturing industry. If microelectronics is going to increase productivity in the services greatly, then it will be difficult to maintain employment in this sector and considerable shifts in skills of service workers will be required. If much typing, filing and accountancy can be automated, then skilled but largely routine operations will greatly decrease their labour requirement. Services will need very highly skilled people and a few reasonably skilled, but very large numbers of those currently employed will become redundant.

Several arguments may be advanced against the above. Efficiency and productivity in the services are not properly defined, for nobody can tell what the proper output of an administrative or sales department or social services unit should be. If the output is indeterminate, then the ratio of input to output—the productivity—also becomes indeterminate.[31] In a variant on Parkinson's Law one might say that the tasks set to a service organisation will grow with the means available and output will match gains in productivity. This cannot happen in a climate of economic stringency, when service output is barely maintained and any productivity gains are used to reduce staff. Microelectronics cannot be regarded as an autonomous determinant—it can be used in a variety of ways. Similar arguments apply to skills. A typist may be made redundant by a word processor, but she might also be used either to increase output or to become a secretary, administrator or manager.

Thus we see yet again that there is no technological determinism —technology does not autonomously force decisions upon people. Yet we must accept that technology exerts considerable pressures and circumscribes the range of possible decisions. These questions will form a major topic in later sections, for the moment we note that the outcome of introducing microelectronics in the office is somewhat indeterminate in terms of both skills and employment. Certainly microelectronics reinforces current pressures for streamlining of office work and endangers the position of the middle range of skilled workers. There will be increased demand for new and very high skills in the office, but the middle ranks may be reduced. Many political battles will be fought here and most white-collar unions have already taken up the cudgels.

(iii) Fears for privacy and freedom

The computer may be regarded as a gigantic card index with very fast access to each card from many, often distant, locations. Viewed in this way the computer can obviously be used to hold much personal information about every citizen and this information can be used against the individual if those holding the information wish to do so and are not effectively prevented from misusing their information. Horrifying pictures can be painted: the secret police recording every move and every utterance of their potential victims and pulling it out of their computer whenever they feel inclined; private organisations similarly holding financial, political or personal details and using them for credit ratings, job enquiries or even blackmail.

Of course the honest citizen has nothing to hide—but nothing to hide from honest people only and, in any case, we all have the need to keep private matters private even if they are perfectly innocent. And of course oppressive regimes and blackmailers of all kinds have flourished since time immemorial, without the aid of as much as a card index, let alone the computer.

We see again the characteristic position of technology, that of enabling people to achieve certain of their wishes. Modern computer technology makes it possible to increase the supervision of individuals, it enables those who seek to hold information to do so. But technology does not force the abuse of information nor is the availability of a computer a necessary condition for abuse. What is so often claimed for technology, that it is neutral, indifferent to what is done with it, happens to be true in a very real sense. Yet in another sense it is not neutral, for only those with the means of obtaining and using a computer can take advantage of the powers it confers and because it enables only certain kinds of activity.

The availability of a major new technology, such as the computer, makes it necessary for the state to enact new measures for the control of possible abuses. There are two very simple principles on which new legislation for the safeguarding of personal information should be based. The first is the principle of 'the need to know'. Personal information must be given only to those who truly and legitimately require it. The doctor treating a patient needs to know his or her medical records, most other people do not need this knowledge. The bank manager considering a loan to a person needs to know his or her financial background, the judge needs the criminal record, the teacher the academic achievements. Details may become complex, but the guiding principle is sound. The second principle is that of 'the right to know'. Each individual should have the right to know what information anybody holds about him or her and should be able to correct errors or demand the erasure of the information if its holder does not need to know it.[32]

There are, of course, special situations. Undoubtedly society has the right to protect itself against crime and to use a properly established police force with proper safeguards for this purpose. The police must use all available relevant technology and indeed the computer and modern telecommunications have proved a tremendous boon in the fight against crime. Unhappily, new technology also produces new types of crime. We now witness computer criminals attempting to misdirect funds; piracy of software despite copyright protection; and the use of tiny transmitters for industrial espionage. A proof of the neutrality of technology, or merely a reminder that new tools require new rules?

ISOLATION

Perhaps the most pervasive yet hardest to articulate fear of modern people is the fear of isolation. Few doubt that modern people feel a sense of isolation which is greater than that felt by their great grandparents. The phenomenon has many facets and many causes, but technology is at least a contributing factor and is identified, recognised and feared as such by many.

Marx spoke of alienation of the factory worker and meant by this particularly the twin facts that the owner of the means of production directed the worker's efforts and appropriated to himself the results of the work done. From this Marx reasoned to the alienation of each man from his fellow men and ultimately from himself. The concept was further developed by several sociologists, among them Durkheim who used the term 'anomie' to describe a sense of estrangement of the individual from society.[33]

We mentioned in Chapter 1 that technology arose out of cooperation of people in tasks and that in this way surpluses were created which could be used to strengthen ritual and thus cooperation. Thus we postulated a benign circle where technology and cooperation gained strength from each other. Since those days of the dawn of civilisation technology has moved a very long way and the benign circle has become a vicious one.

The modern machine is so powerful and so highly developed that it removes much need for cooperation in the accomplishment of tasks. The modern production system, driven by a craving for efficiency, has achieved so much formal organisation in its division of labour that yet again people need hardly work with each other. The power of modern machinery together with the power of modern organisation have combined to isolate the modern worker. His arm is so strengthened by organised technology that the need for immediate human contact and cooperation is much diminished.

Examples abound. The bulldozer has replaced the gang of navvies. A single officer on the bridge of a huge modern ship, peering at the screen of his radar and surrounded by instruments and switches, achieves a multiple of what dozens of sailors, hauling sails and turning wheels, could ever achieve in the past. The engine driver has lost his fireman, the chemical plant or power station are operated virtually single handed, people on production lines cannot and need not communicate with each other.

Communications and cooperation have been removed from the factory floor to supervisory, maintenance, organisational, design and management functions and even there it is threatened by formalisation aided by ubiquitous electronic data processing and data transmission. There are, of course, swings and roundabouts. The international economy has enabled thousands of people to fly around the world and meet their opposite numbers in the four corners of the earth. The telephone has made it possible to talk to people in remote locations. But are the jet setter and telephone addict less isolated than their forebears who worked with their nearest neighbours as their mates?

Nobody in their right minds would envy the crewmen on sailing ships climbing up masts in a howling gale, any more than they would condone child labour in early factories or suggest that the next canal should be dug with pick-axes and shovels again. The fact that we have emerged from an awful past to a better present need not detract from the many imperfections of the present. The fact that technology had eased the burden of Man, need not blind us to the fact that it has also isolated him and that we need to combat this trend of technology.

The power of machines has turned much communal effort into isolated tasks, carried out with the aid of ingenious artefacts rather than fellow human beings. By the very power at his command, *Homo Faber* has become a somewhat solitary worker. Tasks are split up into minute actions, organisations are held together by all-pervading machinery and by formal structures. One would imagine that if this has become the fate of *Homo Faber*, then *Homo Ludens*, the alter ego of Man, must have triumphed and that social forms must have reached new heights of achievement. That this is not so is too self-evident to require any discussion, the only question is why not.

Many would blame the nature of modern technology for this defect too. For technology, mass produced for a mass market and sold to millions of individuals, can do wonders for people's autarkic need and yet can do little or nothing for their cooperative desires. The motor car makes its owner apparently independent of communal transport, the television set makes him self-contained in entertainment, the deep freeze appears to provide autarky in food. The fact that all these devices operate only because there is a vast organisation behind them is immaterial to the individual. The individual sees himself independent of the pack, able to look after himself, not needing anybody. Which is all very well, except that the other side of Man's nature desperately need people and when this need gets the upper hand, they are not there—they have all gone into their own little self-contained high technology boxes. And when the television presenter says 'see you tomorrow' everybody knows this to be untrue and that perhaps nobody will see you tomorrow. Unless, of course, you buy a Japanese home robot who will welcome you home in a pleasant electronic voice of your choice and will switch on your pre-selected television programme.[34]

NOTES

1. *Poor Man's Guardian*, No. 47, 5 May 1832. I am indebted to S. Dey for drawing my attention to this and similar historic passages.
2. Goethe, J. W., *Wilhelm Meisters Wanderjahre*, Buch III.
3. See e.g. Lenin, W. I., 'Thesen zum Referat uber die Taktik der KPR auf dem III Kongress der kommunistischen Internationale', *Ausgewählte Werke*, Band II, s.868, Moskau 1947.
4. Rosenberg, N., *Perspectives on Technology*, Cambridge, Cambridge University Press, 1976, pp. 229–31.
5. For a very brief introduction and further references see Braun, E. and Collingridge, D., *Technology and Survival*, London, Butterworths, 1977.
6. Meadows, D. H., Meadows, D. L., Randers, J. and Behrens, W. W., *The Limits to Growth*, London, Pan Books, 1974.
7. Cole, H. S. D., Freeman, C., Jahoda, M. and Pavitt, K. (eds), *Thinking About the Future*, Brighton, Sussex University Press, 1973.
8. Weinberg, A., 'Science and trans-science', *Minerva*, 10 (1972), pp. 709–22.
9. Ziman, J. M., *Public Knowledge*, Cambridge, Cambridge University Press, 1968.
10. Kuhn, T. S., *The Structure of Scientific Revolutions*, 2nd edn, Chicago, University of Chicago Press, 1970.

11. Collingridge, D., 'The entrenchment of technology—the case of lead petrol additives', *Science and Public Policy*, **6** (1979), p. 332; Collingridge, D. and McEvoy, J., 'The cost effective comparison of controls on environmental lead', *International Journal of Environmental Science*, **16**, No. 3-4 (1981), p. 139; Collingridge, D. and Douglas, J., 'How important are experts?—disjointed incrementalism and the control of environmental lead', *Social Studies of Science*, 1983.
12. Collingridge, D., *The Social Control of Technology*, London, Frances Pinter, 1980.
13. For a very brief discussion and further references see Braun, E., Collingridge, D. and Hinton, K., *Assessment of Technological Decisions*, London, Butterworths, SISCON series, 1978.
14. Williams, J. (ed.), *Carbon Dioxide Climate and Society*, Oxford, Pergamon Press, 1978.
15. Pick, H. J., 'The importance of design for efficient energy utilisation', *Journal of Mechanical Working Technology*, 6 (1982), pp. 253-66.
16. Sutulov, A., *Minerals in World Affairs*, Utah, University of Utah Press, 1973.
17. Braun, E., 'Some thoughts on the use of materials in a technological society', talk given at National Conference of the Institute of Purchasing and Supply, York, September 1976.
18. Interview with E. Schmid, University of Vienna, 1982.
19. See e.g. Braun, E. and Wait, E., *Programmed Problems in Thermodynamics*, London, McGraw-Hill, 1966.
20. See e.g. Foley, G., *The Energy Question*, Harmondsworth, Penguin, 1975; or Chapman, P., *Fuels Paradise*, Harmondsworth, Penguin Books, 1975; or *Deciding About Energy Policy*, London, Council for Science and Society, 1979.
21. Pick, 'The importance of design'.
22. There is a prolific literature on this subject. For a small sample see Collingridge, *The Social Control of Technology*; Foley, *The Energy Question*; Chapman, *Fuels Paradise* and *Deciding About Energy Policy*.
23. Lönnroth, M., Johansson, T. B. and Steen, P., *Solar versus Nuclear— Choosing Energy Futures*, Oxford, Pergamon Press, 1980.
24. For readings on the nuclear debate see e.g. Wynne, B., *Rationality and Ritual: The Windscale Inquiry and Nuclear Decisions in Britain*, London, British Society for the History of Science, 1982. Of the vast literature on the microelectronics debate, the following provide a sample: Barron, I. and Curnow, R., *The Future with Microelectronics*, London, Frances Pinter, 1979; Bessant, J., Bowen, J. A. E., Dickson, K. E. and Marsh, J., *The Impact of Microelectronics: A Review of the Literature*, London, Frances Pinter, 1981; Braun, E. and Macdonald, S., *Revolution in Miniature*, Cambridge, Cambridge University Press, second edition, 1982, chapter 12, pp. 181-218; Braun, E. and Senker, P., *New Technology and Employment*, London, Manpower Services Commission, 1982; Dostal, W., *Beschäftigungswirkungen der Datenverarbeitung*, Stuttgart, Kohlhammer, 1980; Evans, J., *Microelectronics and Employment in Europe in the 1980's—The Trade*

Union Response, Brussels, European Trade Union Institute, 1979; Green, K. and Coombs, R., *The Effects of Microelectronics Technology on Employment Prospects in Tameside*, Farnborough, Gower Press, 1981; Jenkins, C. and Sherman, B., *The Collapse of Work*, London, Eyre Methuen, 1979; Large, P., *The Micro Revolution*, London, Fontana, 1980; Nora, S. and Minc, A., *L'Informatisation de la Société*, Paris, Documentation Française, 1978; Rathenau, G. W. *et al.*, *The Social Impact of Microelectronics*, The Hague, Government Publishing Office, 1980; Schenk, W. *et al.*, *Anwendungen, Verbreitung und Auswirkungen der Mikroelektronik in Oesterreich*, Wien, WIFO, 1981; Sleigh, J., Boatwright, B., Irwin, P. and Stanyon, R., *The Manpower Implications of Microelectronics Technology*, London, HMSO, 1979.

25. See Barron and Curnow, *The Future with Microelectronics*; Bessant *et al.*, *The Impact of Microelectronics*; and Braun and Macdonald, *Revolution in Miniature*.

26. Moseley, R., 'Technical change and employment in the post-war gas industry', *Omega*, 7 (1979), pp. 105–12.

27. Braun and Macdonald, *Revolution in Miniature*, p. 206.

28. For a recent review, with heavy emphasis on technological factors, see Freeman, C., Clark, J. and Soete, L., *Unemployment and Technical Innovation*, London, Frances Pinter, 1982.

29. Braun and Senker, *New Technology and Employment*.

30. Wilkinson, B., *Technical Change and Work Organisation*, Ph.D. thesis, University of Aston in Birmingham, 1981, abridged version published as *The Shopfloor Politics of New Technology*, London, Heinemann, 1983; Braverman, H., *Labour and Monopoly Capital*, New York, Monthly Review Press, 1974.

31. Braun, E., Some Remarks on the Economics of Information, Laxenburg, International Institute of Applied Systems Analysis, Collaborative Paper edited by Vasko, T., *Telecommunications: Some Policy Issues*, pp. 12–19, October 1982.

32. A document produced by the Swedish Ministry of Industry in March 1982 very briefly discusses continued concern in Sweden on vulnerability of data and stresses the right of access to all public information and the severe restrictions on the holding of inaccessible private information on individual citizens. For a recent critical discussion of the legislative position in Britain see Hermann, A. H., 'Three cases of derailed legislation', *Financial Times*, 7 April 1983, p. 40. For a fascinating discussion of human and social problems associated with the computer see Weizenbaum, J., *Computer Power and Human Reason: From Judgment to Calculation*, San Francisco, Freeman & Co., 1976.

33. For a review of the literature see e.g. Wilkinson, B., 'Technology and work satisfaction', MSc thesis, University of Aston in Birmingham, 1979.

34. Garner, R., 'Home robot on the doorstep', *Financial Times*, 2 November 1982, p. 16.

Technology assessment was to be the answer to all fears about technology. The term was coined by the US Congress and the concept suggests that instead of finding harmful effects of a new technology after it had been introduced, such effects would be forecast in extensive studies before the introduction and could thus be avoided. The fear of technology would be sublimated into a positive act of will to avoid unwanted effects by foreseeing them. Foresight would substitute for hindsight; prevention would avoid the need for cure and fear.

The predecessor to Technology Assessment in the US legislature was the Environmental Impact Statement, embodied in the 1969 National Environmental Policy Act. The despoliation of the natural environment by excessive or wrong use of technology was the first of the recent concerns about technology, with precursors in the form of various conservationist movements, and became the first to find expression in new legislation. The 1969 Act, which established a Council on Environmental Quality within the Office of the President, required an annual report to Congress on the state of the environment and demanded that

All agencies of the Federal Government shall include in every recommendation or report on proposals for legislation and other major Federal actions significantly affecting the quality of the human environment, a detailed statement by the responsible official on
  (i) the environmental impact of the proposed action,
 (ii) any adverse environmental effects which cannot be avoided should the proposal be implemented,
(iii) alternatives to the proposed action,
(iv) the relationship between local short-term uses of man's environment and the maintenance and enhancement of long-term productivity, and '
 (v) any irreversible and irretrievable commitments of resources which would be involved in the proposed action should it be implemented.[1]

Complex procedures and rules have gradually grown out of these requirements, which are overseen and enforced by the Environmental Protection Agency. The preparation of Environmental Impact

Statements has become part of the American way of life and no public or private project requiring planning permission or state funding or any kind of state licence, permission or aid can go ahead without an Impact Statement. Whatever the quality of the statements and whatever the difficulties of impartially ascertaining in advance all the environmental impacts a project is likely to have, it must be recognised that the extensive iterative discussions surrounding each statement are bound to have a beneficial effect upon the environment and to reduce its gross abuse.

Even before the National Environmental Policy Act became law, Congressman Emilio Q. Daddario, the then chairman of the House Committee on Science and Astronautics, and his allies were battling for the establishment of a Technology Assessment capability. A bill to establish an Office of Technology Assessment (OTA) was passed by both houses in 1972 and was known as the Technology Assessment Act. The Office was finally established with all its boards and councils at the end of 1973 and ex-Congressman Emilio Daddario became its first director.[2]

Definitions of Technology Assessment abound. We shall quote three—the original definition by Emilio Daddario, an early definition given by the head of the Legislative Reference Service of the Library of Congress and one by Joseph Coates, a well-known practitioner of TA and an erstwhile manager of the National Science Foundation Technology Assessment Program. According to Emilio Daddario,

Technology assessment is a form of policy research which provides a balanced appraisal to the policymaker. Ideally, it is a system to ask the right questions and obtain correct and timely answers. It identifies policy issues, assesses the impact of alternative courses of action and presents findings. It is a method of analysis that systematically appraises the nature, significance, status, and merit of a technological progress.[3]

The definition used in the Congressional Research Service reads as follows:

Technology assessment is the process of taking a purposeful look at the consequences of technological change. It includes the primary cost/benefit balance of short term localised market place economics, but particularly goes beyond these to identify affected parties and unanticipated impacts in as broad and long range fashion as is possible. It is neutral and objective, seeking to enrich the information for management decisions. Both 'good' and 'bad' side effects are investigated since a missed opportunity for benefit may be detrimental to society just as is an unexpected hazard.[4]

According to Joseph F. Coates,

Technology assessment is the name for a class of policy studies which attempt to look at the widest possible scope of impacts in society of the introduction of a new technology or the extension of an established technology in new and different ways. Its goal is to inform the policy process by putting before the decision maker an analyzed set of options, alternatives and consequences. . . .[5]

Whatever the precise wording of the definition, TA is generally understood to mean an attempt to discover all the ramifications and effects which a technology is likely to have when it is in full use at some future date. The study must be interdisciplinary in nature, requiring knowledge and insights from engineering and both the natural and social sciences. Both beneficial and harmful effects need to be described and alternative policies for dealing with the introduction of this or rival technologies should be elaborated. Groups likely to benefit or to be harmed should be identified and the study must be carried out impartially, both with respect to technologies and to social groups.

Technology Assessment is an input to the decision process and the assessment team is not the decision maker. In theory at least, the assessment should be of the form 'if such and such a policy is adopted, such and such is likely to happen and to affect groups A and B in these ways . . .'. It is up to the decision maker to decide which effects are to be achieved and hence which policies are to be adopted. This concept is, of course, an ideal. Nobody pretends that it is possible to foresee all the ramifications of a technology any more than it is possible to foresee all social, economic and technical changes which will occur in the period under discussion. Indeed the chains of causality are so complex that the very concept of cause and effect becomes doubtful. As stressed several times before, events move in trajectories determined by a field of many forces and a constellation of many circumstances. Although omniscience is required to achieve the full idealised goals of Technology Assessment and even the best assessment team is unlikely to be omniscient, it is arguable that a good attempt to approach the ideal is infinitely preferable to doing nothing.

The Office of Technology Assessment is a creation of the US legislature and Congress is its only authorised client. The two main motives for setting up OTA were (i) the need to make technology more socially acceptable and useful; (ii) the overwhelming need of legislators to be better informed about technological and scientific matters for which they were passing legislation and financial appropriations. Ever since the congressional deliberations on whether or not the United States should build a supersonic transport aircraft, Congress felt helpless under a deluge of ill-assorted facts and opinions.[6] Considering that few members of Congress are technically qualified and fewer still are expert in the particular field under discussion at any given moment, their need for pre-digested information and advice becomes overwhelming. The split between assessor and decision maker is neat and unequivocal: OTA assesses, Congress decides.

The concept of Technology Assessment is wider than the need for scientific advice by the legislature and many forms of TA in many different institutional settings are possible and do exist. In fact it has become fashionable to use the term for any attempt at assessing the consequences of technology or decisions about technology. This is unfortunate, as TA should retain the meaning of a broad-based interdisciplinary independent assessment, no matter in what setting.

An assessment can be either problem orientated or technology orientated. The problem posed to TA could, for example, be 'alternative strategies and methods for conserving energy' or 'life extending technologies'. On the other hand the technology to be assessed might be 'coastal effects of offshore energy systems' or 'synthetic liquid fuels development'.[7] The difference between the two types of assessment is not as great in practice as it might seem in theory, because the problem orientated TA must list the available (or likely to become available) technologies for the solution of the problem and analyse their impact, while the technology orientated TA must not only analyse the impact of the given technology but also investigate its actual or potential technological rivals.

A point of contention is the definition of technology. Some authors prefer to include so-called 'soft' or 'social' technologies under the definition and studies such as 'impact of no-fault-automobile insurance' or 'alternative work schedules'[8] have been called Technology Assessments. Although many borderline cases are possible and the above may lie very close to the border, the term technology is used throughout this book in the sense of tangible artefacts, such as tools, instruments, machinery and equipment, used to produce desired effects. In this definition of technology purely organisational arrangements, say social security payments or trading stamps, where the use of physical artefacts is trivial or not of the essence, are not considered as technology. We should also, strictly speaking, distinguish between products of technology, such as clothes, which serve a useful purpose but are not involved in producing any other items, and technology proper, used to produce artefacts or effects, but the borderlines become rather blurred. Machinery and its mode of use, computers and their software, telecommunications equipment and systems, railway systems, handtools, pots and pans, chemicals, roads and bridges, are all examples of technology. Significantly, we include the mode of operations, the software, together with, but not separately from, the hardware in the definition of technology. The essence of our definition is that technology involves hardware as a significant feature of a system designed to achieve some human, primarily material, goal.

Because of the novelty and the enormity of the task, early studies in Technology Assessment were preoccupied with methodology. It was felt that only a systematic methodic approach could lead to reliable results which were not dependent upon the personalities and values of the assessors. The first need was an overall description of all the steps required in a TA; the second need was for methods to be used within each step. Many alternative overall schemes have been suggested and two of these are reproduced in Table 4.1.

The first step in any study must be the definition of the problem. For if one took literally the phrase 'all impacts; first, second and higher order effects', then any TA would approach infinity in extent and in the time required for its completion.

The definition of the problem is normally a matter for negotiation,

Table 4.1. The Mitre/Jones seven-step TA methodology and Joseph Coates's ten elements of TA

MITRE/JONES 1971 7-Step TA Methodology	JOSEPH COATES 1976 10 Elements of TA
1. Define the assessment task	1. Examine problem statements
2. Describe relevant technologies	2. Specify system alternatives
3. Develop state-of-society assumptions	3. Identify possible impacts
4. Identify impact areas	4. Evaluate impacts
5. Make preliminary impact analysis	5. Identify the decision apparatus
6. Identify possible action options	6. Identify action options for decision apparatus
7. Complete impact analysis	7. Identify parties and interests
	8. Identify macro system alternatives (other routes to goal)
	9. Identify exogenous variables or events possibly having effect on 1-8
	10. Conclusions (and recommendations)

Source: From Armstrong, J. E. and Harman, W. W., *Strategies for Conducting Technology Assessments*, Boulder, Westview Press, 1980, p. 7.

formal or informal, between assessor and sponsor. This is only the first point of discussion, generally there are many more and it has been argued that these and other contacts are more fruitful and more important than the final TA reports.[9] The range of technologies to be described, the types of impacts and cross impacts to be investigated and, last but by no means least, the time horizons to be considered must all be decided on. Clearly these decisions also define the cost of the study—or vice versa. Within limits, the study can be arranged to fit the purse of the sponsor. The assessor must remain independent of the sponsor and must not do his bidding, but can only be expected to do as much work as the sponsor is willing to pay for. He who pays the piper must not call the tune, but little money pays but for little music.

The second step consists of a description of the relevant technologies, including rival and supplementary technologies. If we consider off-shore oil operations, then the technologies consist not only of a range of designs for drilling rigs, but also of alternative ways of bringing the oil to shore and storing and distributing it. Apart from rival designs of drilling platforms, one might think of completely submerged systems; apart from pipelines one may think of tankers; and so forth. These technologies require dozens of ancillary technologies which may or may not be developed at the right time: corrosion protection, underwater inspection and maintenance, all weather communication and transport, geological survey methods, and many more.

The first two steps are fairly similar in the two schemes shown in Table 4.1. The third step in the MITRE method requires a description of the state of society. This means that assumptions must be made of how certain social parameters will develop over the period covered by the assessment. How much oil will be required and what price will it command? Will people be willing to work under off-shore oil drilling conditions and what wages and amenities will they require? What oil terminal infrastructure, including schools, housing, medical services, police, legal services and new types of services will be required? Will the installations be too vulnerable in case of international conflict and is such conflict likely? Presumably the Joseph Coates approach subsumes all these types of questions under other steps, including particularly 9.

Steps 4 and 5 in the MITRE scheme and 3 and 4 in the Coates scheme are at the heart of any TA. Out of a very large range of possible examples we shall choose two: the hypothetical case of a shift of British agricultural production and food consumption to a meatless system,[10] and the notorious impact of microelectronics.[11]

The envisaged change to a meatless society would require a momentous change in social attitudes—a different set of state-of-society assumptions. If Britain were to change to a meatless system, the main impacts of such change would be felt by every kind of farmer, by butchers and meat traders, feed-stuff importers and merchants, slaughter houses, the food processing industry, farm machinery producers and many others right down to the population at large as consumers of food.

The effects upon the different groups would vary from total elimination of beef farming to a change to a healthy, though perhaps somewhat monotonous, diet for the population. The impact on the dairy farmers might be particularly interesting. For if milk production were to continue without the slaughter of most males of the species and many of the females, then either a vast exploding and largely unproductive bovine population would have to be supported or new technologies would need to be developed. These would have to make it possible to select mainly female offspring and provide for continued lactation without frequent births. This may be bizarre but is probably quite feasible. In any case it is a good illustrative example of how one technical or social change requires another and how inappropriate the word impact really is, for it is hedged by many conditions. Technology is not something coming out of nowhere and hitting an unsuspecting target, an image which the word impact might conjure up. Technology and its social impacts form a complex web of mutual dependence, which may best be described as 'system changes' instead of impacts.

This becomes even more obvious when we consider the case of microelectronics. To identify the groups upon which this ubiquitous technology might have an impact is difficult to the point of impossibility. Is it an impact if we use a pocket calculator and thus forget our multiplication tables? Is it an impact if we use a push-button telephone which can connect us to any subscriber in the world and can remember those numbers we require most frequently? Or do we count it as an impact only if we lose a job, become de-skilled, or are spied upon by a dictator equipped with a computer? To get away from semantics, the fact is that because microelectronics can be used in a very wide range of consumer and capital goods and makes many new products, processes, and information transactions technically and economically feasible, it will affect practically everybody in some way. Yet the effects will be unevenly distributed. The office worker might be more strongly affected than most, followed by many factory workers, draughtsmen, and so on.

What the effects will be depends on the scenario we choose, and this again is characteristic of TA. The effects of any particular technology depend not only on developments of that technology and its ancillaries and rivals, but also upon developments in the domestic and world political and economic scene. The speed of introduction of microelectronics depends upon availability of investment capital, availability of skills, climate of opinion and, above all, competitive pressures. The employment effect depends upon the general economic position, while the effect upon skills depends, to a degree at least, upon conscious policies.

It is apparent that impacts or effects or system changes, or whatever expression is chosen, depend upon a range of circumstances and forces within and without technology and upon decisions and policies by a variety of decision makers. Hence the emphasis on scenarios—a range of probable effects occuring within a range of probable boundary conditions—and the stress on policy issues.

Point 6 in the MITRE scheme and points 5 and 6 in the Coates scheme emphasise the view that indeed any Technology Assessment is futile unless somebody acts upon the insights gained. The matter goes deeper, for very few things outside the realm of nature, and certainly not matters technological, happen without decisions by people. The essential feature of TA is that it attempts to inform those who make decisions, in the hope that better information will lead to better decisions. Hence the emphasis upon decision makers and upon policy options. Unless a set of policy options is described, an assessment of technology cannot truly claim to be a Technology Assessment.

A Technology Assessment cannot be a substitute for a decision, it must be an input to the decision-making machinery. In the early days of TA and in its embodiment in the Office of Technology Assessment, the decision makers were clearly identified—decisions were made by Congress and its committees. Hence the MITRE scheme is content with a single step to 'identify possible options'. In alternative developments, the concept of TA became transferred to other settings of corporate or public policy and hence the Coates scheme demands identification of the decision apparatus as a separate step. By the time Coates described his methodology, the controversial, adversarial and deeply political nature of impact assessment had become clearer and hence he separated 'parties of interest' from both impact areas and decision apparatus.

Armstrong and Harman, in an analysis of a number of TA studies, identified several features which are common to this class of study. They found what they termed 'three essential functional elements' for all TA and six 'cross-cutting concerns', which run through all the functional elements. The three functional elements are:[12]

Technology description and alternative projections. Defining the current state of the art of the technology and projecting it into the future along feasibly attainable alternative paths.

Impact assessment. Performing a comparative evaluation of the technological alternatives using broadly based criteria that include primary and higher-order social and environmental concerns along with the more conventional elements of economics, technical performance, and legal and institutional considerations.

Policy analysis. Comparing the technological alternatives through consideration of feasible policy options available to the applicable policy community.

The six cross-cutting concerns are:[13]

1. The alternative *whole societal future contexts* within which the technological alternatives are embedded and against which the judgements of impacts and implications are made.
2. The context of specific *societal values* assumed, and the way in which they are accounted for in subelement application.
3. The *uncertainty* of alternative projections of the technology, of the societal and value impacts that result from the technology projections, and of the effectiveness of proposed policy options.
4. The *iteration and sensitivity analysis* used to insure an adequate assessment.
5. The quality of *team interaction* necessary to achieve an integrated and multidisciplinary assessment.
6. *Validation and public participation.*

Apart from the grand schemes of devising major steps or finding common elements, much of the extensive TA literature is concerned with detailed methodology.[14] There is, by now, a degree of consensus that no homogeneous methodology can or should be devised beyond the general features discussed above, but detailed methods must be used to fit individual cases: horses for courses, within a general framework of attitudes, rules and concerns. Among the many methods used, a few stand out as of particular significance.

The first is brain-storming, mostly in an attempt to obtain a comprehensive list of possible impacts and cross-impacts. Under the general rule of 'nothing is too crazy for us', it is hoped to generate enough ideas for the analysts to get on with. But brain-storming is only a ritual acknowledging and taking to an extreme one of the basic features of TA activity (and much other intellectual activity)— the need for interaction between people. What makes this need especially great in TA is the focus on real problems and therefore interdisciplinarity; for real problems just do not fit into neat academic boxes and do not pose neat intellectual puzzles of pure scientists.

Hence the need for interaction between people of different experience and a broad range of inputs, which is superimposed on the common need to clarify ideas and concepts in discussion.

Much has been made of iteration in the context of TA. But the need for iteration is a direct result of dealing with complex issues. Only when the issue has been thoroughly aired, discussed and analysed can we begin to understand it, and that is precisely the time to go back to the beginning and look at the problem afresh. The need for iteration becomes especially urgent when we deal with multiple interactions of many factors, so that the achievement of clarity in the first round would be too much to expect.

The set of techniques known as impact analysis or cross-impact analysis are clearly of importance. A rather nice technique with several variants is the relevance tree, which helps to show how one thing leads to another and how the effects of a technology may branch out. Unfortunately trees grow without apparent feedback loops, so these have to be catered for in different ways. Figure 4.1 shows an example of a relevance tree.

One way of showing the interrelationships between different impacts is by using a cross-impact matrix. This can be either purely qualitative, showing how the occurrence of one event might affect that of another, or it can be quantitative, assigning probabilities of occurrence to the various events.[15]

The first set of techniques is mainly intended for the identification of impacts and impact areas, with cross-impact matrices having some utility in the analysis of the impacts. The next type of techniques, cost–benefit analysis, is entirely aimed at achieving a balanced view of the costs and benefits of identified effects. The method is as widely used as it is criticized. In essence, cost–benefit analysis attempts to assign a price, a monetary value, to each effect, positive for benefit and negative for harm, and to obtain a final net benefit by summing over all partial values and all affected parties.[16] The two chief difficulties are that the assignment of a price is only possible in a market situation and therefore the values obtained are artificial and highly uncertain. To make matters worse, a monetary value has to be assigned to matters which are not easily amenable to this treatment: quality of community life, an old Norman church, a forest and its wildlife, etc. Not many would claim that the values ascertained, often by quite sophisticated methods, have absolute validity, but as a method of comparing the relative benefits attainable by different solutions to a given problem, cost–benefit analysis has a lot to commend it. The best known example of the extensive use of cost-benefit analysis is the Royal Commission on the siting of

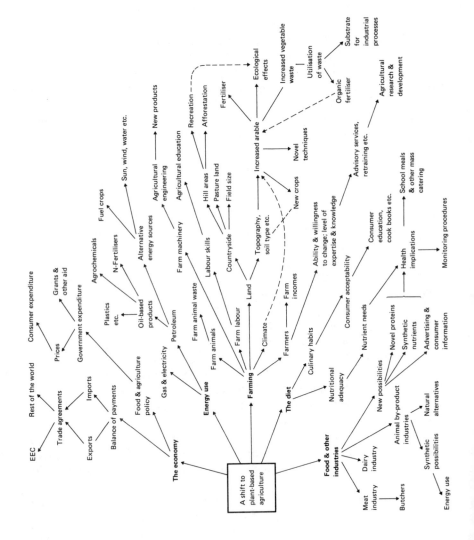

Fig. 4.1. Relevance tree for assessment of impacts of vegetarian agriculture. *Source*: Thompson, S., 'The Potential for and Limitations of a Shift from Animal-Based to Plant-Based Agriculture and Food Production in England and Wales.' Ph.D. thesis, University of Aston in Birmingham, 1979.

the third London airport.[17] Ironically, the British Government brushed aside the committee's recommendation and embarked upon a programme of extension to the existing airports instead of building a new one. Travellers through Heathrow may be excused for thinking that the committee made more sense than the Government—but each acted out its legitimate respective role of assessor and decision maker and we cannot complain about that, though we can, if we wish, complain about the quality of the decision.

Even this very brief and incomplete list of TA methods must make some mention of simulation models. In this age of the computer, an enormous amount of effort has been put into programming of computer models of large systems. The essential idea is that relationships between magnitudes that can be represented by a number or a probability, can be simulated by a system of equations which the computer can solve for the full range of the variables. The variables may be numbers of certain machines, degree of pollution, number of people with a certain qualification, gross national product, probability of completion of a technical development, or any other variable which can be expressed in numerical terms. The results of the computer simulations are only as good as the assumptions and relationships incorporated into the model, but being able to observe the effect of changes in individual variables is a great aid to thinking out consequences of small changes in large systems. The computer is an excellent complement to human thought—no inspiration, judgement, or intuition, but excellent at solving lots of equations.[18]

Many of the effects which TA attempts to assess depend on future developments of technology and of society. Unless we are dealing with historical assessments,[19] which are interesting and didactically useful but of limited immediate utility to the decision maker, TA is inextricably bound up with forecasting. Forecasts attempt to obtain scenarios of future developments of both technology and society. Technology Assessment attempts to use the forecasts as a basis of analysis for future effects of technology. Take energy supplies as an example. If an assessment is to be made of the breeder reactor and its role in the future energy supply system of a given country, we need forecasts of several kinds.[20] First, we need to know the total likely demand for electricity and any shortfall in the supply of more traditional sources of power. Secondly, we need to know the likely price of uranium and of fuel processing technologies as well as the ultimate construction costs for breeder reactors. This in turn depends strongly upon interest rates and upon international exchange rates on the one hand, and upon learning from previous experience and technological developments on the other. Thirdly, we need to know how quickly and how effectively rival technologies, such as solar power or nuclear fusion, will be able to compete with breeder reactors. Fourthly, we need to know how acceptable the risks associated with the handling of radioactive materials will be by the time the breeder programme reaches full stride. To say that Technology Assessment depends upon forecasts and needs forecasts in order to carry out its analysis, is understating the intimacy of the relationship. Forecasting is part and parcel of most Technology Assessments and forecasting methods belong firmly in the methodological armoury of TA.

It is indeed arguable that forecasting is only useful in the context of policy analysis. For only 'if–then' types of forecasts, where 'if' is a policy and 'then' a likely outcome, provide useful guidance for actions.[21] Because of the notoriously unreliable nature of forecasts, any kind which is unrelated to policy options appears little more than an extravagant intellectual game.

Among the many forecasting methods used, two appear particularly common and fruitful. One is trend extrapolation and analysis, the other is the formation of a consensus of expert opinion, known as the Delphi method. Trend extrapolation can use sophisticated statistical methods to obtain the best fit of existing data to a curve extrapolated into the future. Great care must be exercised about new factors coming in to influence the past trend and about the point at which the curve will turn or flatten. Fast growth almost inevitably comes to an end—the question is when. Similarly, technical developments tend to follow certain trends and these too tend to flatten off at some point which is difficult to foresee. Well-known examples are the power output per unit of weight of internal combustion engines or the number of logic gates per unit of silicon area in microelectronics. The tailing off of the trend is associated with the law of diminishing returns—the further along the line you are the more expensive each little improvement tends to become. Unless, of course, somebody thinks of a neat way of sidestepping the main obstacle to improvement and a new trend starts. So in car engines it is far easier to introduce a turbo-charger than to design a new engine, if increased power with only slight loss of fuel economy is desired. Because of the law of diminishing returns, the development of a new car engine is now so expensive that even large manufacturers cannot afford it very often. Similarly, further improvements in aero engines have become so hard to achieve that even the giant arch-rivals, Rolls-Royce and Pratt & Whitney, are said to have decided to cooperate on the next generation of jet engines. Airframe manufacturers too try to spread the cost and risk of airliner development by extensive sub-contracting.[22] Technical developments at their extreme limit can make nonsense of competition, whether they can make economic sense must remain a moot point. It would, however, take a brave forecaster to predict the point at which something like major aero engine developments would stop.

Some forecasts based on trend extrapolation have achieved notoriety by their inaccuracy. All forecasts for electricity consumption in Britain have been widely off the mark, with a resultant overcapacity in generating plant.[23] Similarly, forecasts for growth in the number of airline passengers, especially business travellers, have proved wrong and this has been a contributing factor to the current virtual bankruptcy of most airlines.[24] In both cases, extraneous circumstances changed in a way not foreseen, and probably not foreseeable, and a changed trend inevitably fooled the forecasters.

The Delphi method is based on the slightly forced formation of a postal consensus by a group of experts on the likelihood and timing of certain events. A typical question might be 'when do you expect a non-toxic variety of lupins to become commercially available?'[25] The anonymous answers in the first round might range from, say, two years to never. The panel of experts is sent information on the weight of opinion given in the first round and are asked whether and how they wish to modify their first-round replies. A third round might be entered into, although with each round the number of respondents tends to decrease. In the end, a statistically treated result is obtained, assigning probabilities to the events. Whether the statistically weighted opinion of experts is a sound basis for forecasting events must be questionable, but so must be everything to do with the unknowable future. The Delphi method is a ritualised form of obtaining expert opinion, and expert opinion in all forms plays a very major role in Technology Assessment. The assessor is bound to obtain information from the expert and opinion is often thrown in with the information in a mix which is hard to disentangle.

Many of the definitions and all the expectations of Technology Assessment include the adjectives 'unbiased', 'objective', 'neutral' in relation to the nouns 'assessment', 'description', 'analysis'. It is the expectation of neutrality, transferred from the realm of the natural sciences, which forms a key problem area of the assessment process. The objectivity of the natural sciences is based on the repeatability of experiments; the same result can be obtained by any number of observers. The second cornerstone of scientific objectivity is consensus on the theoretical interpretation of observed façts. Theory may be hotly and even passionately disputed at some stage of its development, but by the time it becomes incorporated into the canon of knowledge and before it is displaced from it by a newer theory, all competent members of the particular discipline hold the theory to be correct. Thus objectively observed facts are objectively interpreted by correct theory—at least during periods of stability in the particular branch of science. Disagreement becomes tantamount to incompetence. Only periods of turbulence lead to factions and accusations of bias.[26] Bias of the body-scientific in its choice of problems and its organisation is quite another matter, which need not concern us at the moment.

The position of the unbiased and disinterested observer is the natural ideal for the technology assessor. Data and information are collected from all sources and, in theory, all assessors should end up with the same data base. The data should then be analysed, not perhaps by theory quite as well established as a scientific theory, but nevertheless in a way on which competent and reasonable people would agree. The policy options are then neatly laid out in the form 'if you wish to achieve result A, you should adopt policy Y'. The policy maker then exerts choice, bias, political interests— all of which are prerogatives of the decision maker but taboos to the analyst.

Happily or unhappily, real life is not as simple. The data base for a Technology Assessment does not consist of laboratory observations obtained from immutable nature according to strict rules. Instead, it consists of a complex web of fact, conjecture and opinion obtained from a wide range of written and oral sources. Different assessors have access to different sources and often, alas all too often, the data available are but the tip of the iceberg of secret information; secret because of commercial or political or personal confidentiality or simply out of the convenience of the cabal. Such information is largely inaccessible to the analyst and often even the knowledge of its existence remains hidden. We must accept that the data base for a Technology Assessment will be incomplete and variable according to the methods and possibilities of collection and selection by different groups of analysts.

If the data base is uncertain, the analysis, forecasts and conclusions are positively fraught with difficulty. To select and evaluate written material, or oral evidence, to make assumptions necessary for the working of a computer model or a trend extrapolation, even selecting interviewees and posing questions to them, are activities subject to wide variability. With all the care in the world and all the methodological armoury of the social sciences it is impossible to ensure that the results and conclusions of the same assessment task carried out by different teams will be the same. Applaud as we must the ideal of objectivity in TA, an element of subjectivity unavoidably intrudes.

Accepting the inevitable, ways need to be devised to maintain the utility of TA in the face of subjective judgements. Two main ways and many variants have been suggested to deal with the problem. The first solution lies in the composition and institutional location of the assessment team, while the second demands plurality of assessments.

The composition of the assessment team must, of course, be such as to incorporate the essential skills, knowledge and experience required for the task. But beyond the necessary professional composition of the team it is necessary that it should be balanced in its values and have no particular allegiance to any interests involved in the assessment. This is a tall order, for a balance of values within the team means considerable differences of opinion and although controversy can be creative, too much of it is paralysing. The role of the team leader as conciliator is acceptable, that of lion tamer is not. At best, the balance will only be a rough and ready one, but this, coupled with the highest principles of scientific integrity and conscious objectivity should provide sufficient safeguards for a reasonably objective outcome, provided the condition of non-allegiance is fulfilled.

On this model of Technology Assessment by a balanced objective team, an independent institutional setting is of the essence. Unlike the dustman Doolittle, the assessors must be able to afford moral values and must neither incur nor fear to incur sanctions for any opinions expressed. It has been suggested elsewhere that universities are suitable institutions for the location of TA activities,[27] but of course there are many other institutional frameworks which ensure independence from interests involved in the assessment and where the free expression of opinion is not an act of heroism but the normal way of life.

The pluralistic or adversary model of TA takes the view that differences of values and allegiances to causes should not be suppressed and fought out behind the closed doors of the assessment team, but should instead be brought out into the open. If there is a plurality of opinion and, more to the point, a plurality of interests, let there be a plurality of assessments. On this model, the single 'objective' assessment will be replaced by a number of assessments carried out on behalf of each interested party. Two problems immediately spring to mind. First, the interested parties have to be identified and this identification may not always be easy and does indeed form part of the assessment process. Secondly, the number of assessment teams may have to be considerable, the amount of effort and costs involved very large, total coverage uncertain and the results rather unwieldy. If the number of affected parties is small and easily identifiable, the pluralistic model may have something to commend it, but in many cases it may be impractical. A variant of two independent assessments may have some virtue.

Assuming that two or more assessments are produced, there are at least two ways of dealing with them. The easy way is to supply all the assessments to the decision maker and let the decision-making apparatus digest the multiple assessments instead of a single one. The other way is to insert an additional assessment stage, a kind of court, which will pass judgement on the assessments. With the court of law clearly in mind, the advocates of this procedure imagine that each assessment team would present its case, in the manner of counsel, and the learned judges would decide which is the better case. In fairness, the 'assessments' to be brought before the court would not be comprehensive in any sense, they would consist of pleas for a case, and the assessment process would be carried out by the court on the basis of these pleas. The 'judges' would perhaps be suitably qualified and respected scientists.[28]

The idea of a court as an intermediary between decision maker and interested parties is the old one commonly used in British committees of inquiry, as for example in controversial planning applications. Ever since the famous Windscale inquiry into planning permission for the erection of a nuclear fuel reprocessing plant, it has often been assumed that such judicial inquiry fulfils the same role as a Technology Assessment.[29] This is confusing. The judicial way can be brought into harmony with the idea of TA only if we substitute an assessment team for the court and oblige it to listen, preferably in public, to the pleas made by different parties. Thus the assessors would remain assessors but would also fulfil the role of the judicial inquiry in being obliged to consider all pleas. This method would seem to achieve the best of both worlds: a single balanced team assessment with the assurance that all interested parties were able to state their respective cases. The assessors remain analysts and marshallers of information, and do not usurp the role of judges. It is for the decision maker to judge. We must not confuse policy analysis with politics proper.

The simple judicial analogy is not a good model for any decisions other than 'yes–no', 'guilty–not guilty'. Technological issues rarely are of this nature and call for sophisticated compromises. Such matters are best settled out of court.

The advantages of a competent interdisciplinary team, backed up by adequate library and computer facilities and equipped with membership of various informal information networks are overwhelming. Knowledge travels best 'on the hoof' and scientific knowledge is no exception to this, thus membership of scientific networks is vital. If the team operates in an independent institution with reasonable safeguards against partiality and undue pressures and is properly balanced and well led, then the only objection that could be raised against Technology Assessments produced in this way is the lack of representation of specific interests. In fact the disinterested nature of the team makes it all the more necessary for special pleading to be heard. Thus the team ought to be obliged to receive written submissions and perhaps also conduct public hearings. To this purpose the committee would need to acquire additional expertise in rules of procedure, but this should be no insuperable obstacle. In fact it should be easier for non-expert groups to make representations and be heard and considered by a slightly informal team of assessors rather than by somewhat forbidding judicial procedures.

The combination of a professional assessment team with an obligation to hear all interested parties who wish to be heard appears to offer the best of both worlds—impartiality as well as open conflict of interest. The proper resolution of these conflicts is not part of the assessment process. They merely need to be brought into the open for consideration and finally resolved in the decision making process. The latter is a political step which is designed to cope with conflict—the assessment step is not.

The original institution for providing Technology Assessments was the Office of Technology Assessment and its task was clear cut: to provide TA studies at the request and for the benefit of the US Congress.[30] It is remarkable that no other legislative body has emulated the US model, despite a great deal of discussion in several countries.[31] The reason probably is that the division between legislature and executive is much stricter in the United States than elsewhere. Consequently, outside the United States the governing parties tend to leave the assessment of technology to the executive and the opposition cannot do much about it and, in any case, thinks of itself as the future governing party. The executive, on the other hand, carries out a variety of departmental assessments in a variety of forms. Interdepartmental rivalries effectively prevent the creation of centralised institutions for Technology Assessment at the level of the executive.

Government departments have to form opinions about technologies for several reasons, depending upon their departmental responsibilities. Obvious areas of concern to different ministries are defence, transport, telecommunications, manufacturing technology, safety at work, skill requirements, competitiveness in foreign trade, energy supplies, health care. Ministries have respective responsibilities for planning, R & D, regulations, stimulation and support schemes and infrastructural arrangements. In many of these tasks technology plays anything from a major to a subsidiary role and ministers have traditionally received advice from civil servants, aided by advisory committees, commissioned consultancy reports and a variety of more or less public inquiries.[32] The latter operate mainly in planning cases or after major disasters. A specifically British feature is the occasional appointment of a Royal Commission to conduct an inquiry into issues of major public concern. The Royal Commission on the Environment is the only one of these commissions which has a permanent existence and reports from time to time on environmental issues.

Many of these government advisory arrangements pre-date the idea of Technology Assessment considerably and have evolved over a long period of time. US government interest in technology is said to date from the need to intervene in the many boiler explosions which occurred on early steamships in about 1830.[33] Controls over industrial emissions in Britain go back to the alkali act of 1863 and even that had precursors.[34] Many of these advisory arrangements, particularly in Britain, show clear signs of evolutionary development with infrequent major revision, such as occasioned by the so-called Rothschild report in 1972.[35]

It has become customary to lump all these advisory arrangements together under the term Technology Assessment.[36] This is unfortunate, for it creates the impression that technology is being adequately assessed, controlled and supported. In reality most of the studies carried out are on a small scale and with a relatively narrow brief—a far cry from the ideal of the comprehensive assessment called for by the protagonists of TA. This, coupled with the fact that the almost Messianic belief in the powers of TA, which was part of its early sales campaign, became rapidly disillusioned with the publication of the first far from perfect TA reports, has led to a 'business as usual' atmosphere. Those in power feel that the advice they obtain is as good as it can and need be, those who mistrust technology feel that their mistrust is as amply justified as it has ever been.

Apart from central government, advice on matters technological is needed by local government, quasi-governmental agencies and, of course, industrial corporations. The latter need comprehensive information as part of their product and investment planning process, but also as a hedge against regulatory or societal developments which might adversely affect their business. Clearly an electricity supply corporation needs to be, and generally is, as well informed as government on the prospects for energy supplies. Similarly, the pharmaceutical industry needs to know what scientific developments are just beyond the horizon and also what additional safeguards on the introduction of drugs are likely to be imposed.[37]

Who, then, is to carry out Technology Assessments, who is to use them and who is to pay for them. Clearly, industrial firms will and must do their own in-house or consultant assessments and evaluations. These are paid for by the firm and are unashamedly partisan —carried out in the best interest of the firm. The best interest may be served by an 'objective' study, but that is for the firm to judge. This class of assessments should be called industrial evaluation of technology and not TA in the strict sense.

Guardians of the public interest, which is, after all, what governments and parliaments are supposed to be, have to proceed more cautiously. Clearly, government departments have to continue their various activities for the assessment of individual projects to inform individual decisions. But much more is needed. The more important technologies and larger decisions should be subjected to proper Technology Assessments, carried out by competent independent institutions and financed either directly or indirectly by government. The indirect funding would have the advantage of a buffer organisation between government departments and the TA institutions. In this way inter-departmental rivalries and sectoral interests could be by-passed and the buffer organisation itself could have some competence in deciding what Technology Assessments to commission. In the United States, the National Science Foundation has acquired considerable experience in this matter.[38]

In this way competence in TA could be gradually increased and the results would be accessible to all—executive, legislature, pressure groups and public at large. The extra departmental and extra government nature of the Technology Assessment institutions and their funding organisation would ensure maximum independence, minimum political interference and, hopefully, maximum utilisation of results.

It has been said that the results of TA are as yet disappointing not only because the reports are not as good as they might be, but also because the reports are not widely read. The counter-argument has been that the process of consultations, conversations, interviews and gathering of material has itself been very influential in policy formation.[39] Perhaps in TA it is more important to travel hopefully than to arrive.

The degree of influence of a TA report is rather like the degree of market penetration of a new product. If the report says what everybody wishes to hear it is likely to appear influential and its implied recommendations—or obviously favourable policy options —will be widely discussed and probably implemented. If on the other hand the report deals with unpopular or obscure matters, it is likely to fall upon deaf ears. An almost classical example of the popular type of report is one produced by the Swedish Futures Secretariat on Solar versus Nuclear energy. To cut a long story short, it appears to show that Sweden could manage without nuclear energy and use renewable sources as its mainstay of energy supplies.[40] Not surprisingly, this has become government policy. On the other hand, an equally widely discussed report on the siting of a third London airport,[41] has remained so much dead paper because of local opposition groups and shortfalls in the growth of numbers of air passengers.

Technology Assessment is an ideal. Strive as we may and must, the best we can achieve is a gradual improvement in our present understanding of the effects of technology and through improved understanding better control. Collingridge has argued that because forecasts are bound to prove erroneous, one useful criterion for the assessment of technology is its flexibility.[42] As errors of forecasting become apparent, it should be easy to correct them by adjustment of the technology. If a technology is inflexible, as for example the breeder reactor, this should count heavily against it in any assessment.[43] On this argument small technologies are preferable to large and small steps better than leaps in the dark. The argument is certainly a convincing one, but can provide only one criterion among many and does not reduce the need to develop the art of Technology Assessment very much further.

The requirement of flexibility does, however, impose the need to regard any Technology Assessment as temporary, subject to review whenever circumstances change in any important respect. Thus the process of assessment becomes indeed a process rather than a once and for all monumental statement. The frequency of reassessment will depend both on the rate of relevant change and on the importance of the technology or problem to any particular decision maker.

Some technologies are subject to highly centralised decisions and the model of 'proponent of technology–assessor–affected parties–

decision maker', implied in the original concept of TA, holds. Examples are breeder reactors or supersonic aircraft. The proponent is an obviously interested industry wishing to produce and sell the technology, provided the government subsidises development costs and licenses the product. The government is therefore a major party to the decision. The other decision makers are very small in number and very expert, in our cases the electricity utilities and the airlines. The affected parties need to be identified by the assessment but may include wide sections of the population.

Microelectronics falls at the opposite end of the spectrum and decisions about it are decentralised and diffuse. Large numbers of manufacturers compete for the favours of, sometimes, millions of potential consumers and each of these must be regarded as a decision maker. Technology Assessments on the impact of micro-electronics or some aspects of microelectronics cannot therefore be regarded as advice to a very important section of the decision makers. The only people who might be influenced by assessments are public bodies, who may consider support policies for the manu-facture or diffusion of some microelectronic products or systems. Thus TA becomes an aid to the policy maker who has to decide about support policies, or perhaps about procurement or regulations, but the TA cannot influence the very core of existence of a diffuse technology. Thus the nature of the decision-making process has a considerable bearing upon the possible influence of TA and about the questions which should be posed to it.

NOTES

1. Porter, A. L., Rossini, F. A., Carpenter, R. A. and Roper, G., *A Guidebook for Technology Assessment and Impact Analysis*, New York, North Holland, 1980, p. 29.
2. Porter *et al.*, *A Guidebook for Technology Assessment*, p. 34.
3. Hetman, F., *Society and the Assessment of Technology*, Paris, OECD, 1973, p. 54.
4. Hetman, *Society and the Assessment of Technology*, p. 57.
5. Armstrong, J. E. and Harman, W. W., *Strategies for Conducting Technology Assessments*, Boulder, Westview Press, 1980, p. 1.
6. Tarr, J. A. (ed.), *Retrospective Technology Assessment*, San Francisco Press, 1977, p. 2.
7. Armstrong and Harman, *Strategies for Conducting Technology Assessments*, pp. 99–104.
8. Ibid.
9. Smith, M. Y., 'The experience of the UK Programmes Analysis Unit',

Impact Assessment Bulletin, **1**, No. 3 (1982), pp. 40–53; Smith, M. Y., 'Uses of assessments of technology in the UK Central Government: the experience of Programmes Analysis Unit', Ph.D. thesis, University of Aston in Birmingham, forthcoming.

10. Thompson, S., 'The potential for and limitations of a shift from animal-based to plant-based agriculture and food production in England and Wales', Ph.D. thesis, University of Aston in Birmingham, 1979; Thompson, S. and Braun, E., 'Cropping for plant-based agriculture', *Food Policy*, May 1978, pp. 147–9; Stanley, R., 'The future role of novel food sources in the UK', Ph.D. thesis, University of Aston in Birmingham, 1980; Blackman, C. R., 'Strategies for agricultural change and the UK balance of payments', Ph.D. thesis, University of Aston in Birmingham, 1981.

11. Bessant, J., Bowen, J. A. E., Dickson, K. E. and Marsh, J., *The Impact of Microelectronics—a Review of the Literature*, London, Frances Pinter, 1981; Braun, E. and Macdonald, S., *Revolution in Miniature*, Cambridge University Press, second edition, 1982, chapter 12, pp. 181–218; Evans, J., *Microelectronics and Employment in Europe in the 1980's—The Trade Union Response*, Brussels, European Trade Union Institute, 1979; Nora, S. and Minc, A., *L'Informatisation de la Société*, Paris, Documentation Française, 1978; Schenk, W. *et al.*, *Anwendungen, Verbreitung und Auswirkungen der Mikroelektronik in Oesterreich*, Wien, WIFO, 1981; Sleigh, J. *et al.*, *The Manpower Implications of Microelectronics Technology*, London, HMSO, 1979; Rathenau, G. W. *et al.*, *The Social Impact of Microelectronics*, The Hague, Government Publishing Office, 1980.

12. Armstrong and Harman, *Strategies for Conducting Technology Assessments*, pp. 11–12.

13. Ibid., p. 57.

14. Porter *et al.*, *A Guidebook for Technology Assessment*; Hetman, *Society and the Assessment of Technology*; Armstrong and Harman, *Strategies for Conducting Technology Assessments*; and Medford, R. D., *Environmental Harassment or Technology Assessment*, Amsterdam, Elsevier, 1973.

15. Porter *et al.*, *A Guidebook for Technology Assessment*, pp. 190–5.

16. Layard, R., *Cost Benefit Analysis—Selected Readings*, Harmondsworth, Penguin, 1976; Newton, T., *Cost Benefit Analysis and Public Administration*, London, Allen & Unwin, 1972.

17. Commission on the Third London Airport (chairman: The Hon. Mr Justice Roskill), *Report*, London, HMSO, 1971.

18. Meadows, D. H. *et al.*, *The Limits to Growth*, London, Pan Books, 1974; Bremer S. A., *Simulated Worlds*, Princeton University Press, 1977.

19. Tarr, *Retrospective Technology Assessment*.

20. The breeder reactor is a particular type of nuclear reactor which uses uranium very much more efficiently than other reactors but which exists, as yet, in prototype form only. For a policy discussion of breeder reactor problems see e.g. Collingridge, D., *The Social Control of Technology*, London, Frances Pinter, 1980, pp. 104–9.

21. Encel, S., Marstrand, P. K. and Page, W. (eds), *The Art of Anticipation*, London, Martin Robertson, 1975.

22. Rosenberg, N., 'US Technological Leadership and Foreign Competition: *'De te fabula narratur'?'*, paper presented to a National Academy of Sciences Panel on Advanced Technology Competition and the International Allies, December 1981.
23. For current forecasts and current use of electric power the reader is referred to the annual reports by the Electricity Council.
24. See the Roskill *Report*, appendix 6, pp. 189–94 and *Report of the Public Inquiry into the Building of a Second Terminal at Gatwick Airport* (Inspector His Honour John Newey), 29 January to 11 July 1980, London, HMSO.
25. Stanley, 'The future role of novel food sources in the UK'.
26. Kuhn, T. S., *The Structure of Scientific Revolutions*, 2nd ed., Chicago, University of Chicago Press, 1970.
27. Braun, E., 'Technology Assessment and the role of the universities', *Science and Public Policy*, 4 (1977), pp. 224–9.
28. Dierkes, M. and von Thienen, V., 'Science Court—ein Ausweg aus der Krise?', *Wirtschaft und Wissenschaft*, 4 (1977), pp. 2–14.
29. Parker, R. J., *The Windscale Inquiry: Report*, London, HMSO, 1978; and O'Riordan, T., 'Assessing the environmental consequences of energy development in the UK (Windscale Inquiry, Report of the Commission on Energy and Environment)', paper presented at International Symposium on the Role of Technology Assessment in the Decision-Making Process, Umweltbundesamt, Bonn, October 1982; Wynne, B., *Rationality and Ritual: The Windscale Inquiry and Nuclear Decisions in Britain*, London, British Society for History of Science, 1982.
30. Office of Technology Assessment, *Annual Report to the Congress*, Washington, 1981.
31. Böhret, C. and Franz, P., *Technologiefolgenabschätzung*, Frankfurt, Campus, 1982.
32. Gummett, P., *Scientists in Whitehall*, London, Macmillan, 1981.
33. Porter *et al.*, *A Guidebook for Technology Assessment*, p. 30.
34. Macleod, R. M., 'The Alkali Act Administration 1863–84: the emergence of the civil scientist', *Victorian Studies*, 9 (1966), pp. 85–112; and Te Brake, W. H., 'Air pollution and fuel crises in pre-industrial London 1250–1650', *Technology and Culture*, 16 (1975), pp. 337–59.
35. *Framework for Government Research and Development*, Cmnd 5046, London, HMSO, July 1972.
36. Böhret and Franz, *Technologiefolgenabschätzung*.
37. Steward, F. and Wibberley, G., 'Drug innovation—what is slowing it down', *Nature*, 284 (1980), p. 118; and Steward, F. and Wibberley, G., *Drug Innovation and Public Policy*, London, Croom Helm, forthcoming.
38. Menkes, J., 'The contribution of technology assessment to the decision-making process', Bonn, International Symposium on the Role of Technology Assessment in the Decision-Making Process, October 1982.
39. Smith, 'The experience of the UK Programmes Analysis Unit'.
40. Lönnroth, M., Johannsson, T. B., Steen, P., *Solar versus Nuclear*, Oxford, Pergamon Press, 1980.

41. The Roskill *Report*.
42. Collingridge, D., *The Social Control of Technology*, London, Frances Pinter, 1980.
43. Collingridge, D. and Braun, E., 'A new approach to Technology Assessment —the case of the breeder reactor', paper presented at International Symposium on the Role of Technology Assessment in the Decision-Making Process, Umweltbundesamt, Bonn, October 1982.

5 Government Policies for Technology

INTRODUCTION AND DEFINITIONS

The current interest in technology policy arises out of a paradoxical twin concern: on the one hand the wish by each nation not to be left behind in the race for technology-based prosperity and on the other hand multi-faceted fears about undesirable social and environmental consequences of technology. The belief that better technology means higher economic efficiency and international competitive advantage is a strong driving force behind attempts by most governments to support, foster and accelerate technological innovation. The same recognition also forces governments into a range of policies for infrastructural support of technology in fields as diverse as education and training, telecommunications, fundamental research, patent laws and technical standards. The fears about technology, on the other hand, force governments into a range of regulatory activities, and sometimes also into political battles for their procurement and support efforts.

There is no doubt that in a market economy the major force affecting technology is the market. Nevertheless, public bodies feel compelled to intervene in a regulatory mode in order to avoid damage to health and the environment which unbridled technology might cause, and in a supportive mode to enhance and accelerate the perceived advantages of technology. Public bodies act also as major users of technology in fields as diverse as administration, defence, public transport and communications.

In a general sense, technology policy encompasses all the forces, influences, institutions and measures which determine the emergence and application of technology and thereby shape many features of society. In a strict sense, a policy is a general decision about the direction in which some process should be influenced; 'a decision about future decisions' in Kenneth Boulding's parlance.[1] For epistemological purposes it is more useful to include forces and institutions causing or implementing policy decisions, together with active and passive policy decisions, under the same umbrella term of technology policy. If we wish to understand the multiple relationships between society and technology, we need to explore the forces, institutions and influences as much as actual policies. We can neither understand policies nor their origins and constraints unless we study them in the wider context suggested by our definition of technology policy.

The nature of the forces influencing technology covers a wide range, from the internal logic of pure science at one extreme to power politics at the other. There is, of course, no hope for a complete, comprehensive, teleological understanding of technology policy in the broad sense; but if complete understanding must remain elusive there is great merit in partial understanding and illumination —understanding to enable better policies to be formulated and implemented; illumination to obtain a clearer vision of the bewildering world we live in. Our aim is to explore the forces which shape technology in order to better comprehend the path taken by technological society. As one aspect of this task, we have to explore how government and its agents can and do influence technology.

To progress from the almost metaphysical generality of our previous definition of technology policy to an exposition of deliberate government policies for technology, we have to define technology policy at an operational level. At that level, we define technology policy as the totality of measures by private or public bodies designed to control the creation, application and use of technology. For the purposes of the present chapter we can narrow the definition down by excluding private bodies and concentrating on governmental organisations. We also have to broaden the definition, for many government policies designed to attain a variety of goals— perhaps regional development or a healthy balance of trade—may directly and substantially influence technological developments. Insubstantial or very remote influences still have to be disregarded, as otherwise the problem becomes unbounded. Our final operational definition of technology policy for this chapter therefore reads: 'Technology Policy is the totality of measures taken by government and its agents which directly control the creation, application and use of technology'.

Perhaps we should emphasise here once again that we take technology in the sense of tangible artefacts used to produce desired effects. Strictly, we should distinguish between technology as a means of making things, and products of technology which serve a purpose other than production. At the simplest level a skirt is a product of technology and a frying pan a means of production. For our purposes the distinction is not always helpful and we shall mostly blur it, except when dealing with the difference between investment and consumer goods.

MEASURES AVAILABLE TO GOVERNMENT

The totality of measures which governments can and do take to influence technology is infinitely varied. However, the infinite variety yields readily to classification and some insight can be obtained in this way.[2] Government policies can be classified according to three criteria: (i) aim of policy; (ii) type of measures; (iii) target of measures.

Apart from very specific policy objectives there is now a recognised range of technological infrastructural activities which governments accept as their legitimate responsibilities. These include education at all levels, transport and communications, technical standards and regulations about the use of technology, fundamental research, and a legal framework regulating monopoly, competition, professional activities and many more.

In recent years two major convictions have gained considerable strength. On the one hand it is widely believed that economic growth can only be regained and sustained with the aid of a high rate of technological innovation and on the other hand it is thought that the required rate of successful innovation cannot be achieved without government support. The corollary of this twin conviction is that any industrial country with an insufficient rate of innovation is bound to suffer relative economic decline and thus government aid to innovation is necessary for a country to hold its own in international competition. We might indeed speak of a new mercantilism, whereby governments no longer attempt to increase their country's share of gold, the old symbol of wealth, but rather their share of the new symbol of affluence: advanced technology.

Thus a specific policy objective of most governments of industrial nations is the stimulation of technological innovation and this objective shall serve as our first example of a technology policy. The first question we must ask is what government can do to achieve this aim. We may distinguish at least ten different types of measures which can be taken towards this general aim. Table 5.1 lists the types of measures and gives some examples of possible activities.

The different types of measure can be targeted in different ways. If the general objective is the stimulation of innovation as in our example, then we may distinguish four different target areas for policy measures: (i) general ambience, i.e. the economic, social and political atmosphere in which innovation takes place; (ii) industry in general, i.e. policy measures to strengthen industrial performance and industrial investment or measures to strengthen some specific industrial sectors; (iii) innovation in general, i.e. measures aimed at

Table 5.1. Types of policy measures aimed at stimulating technological innovation

Type of measure	Examples
Financial	Grants, loans, subsidies, financial sharing arrangements, loans and gifts of equipment, provision of free services, provision of buildings
Taxation	Company, personal, indirect and payroll taxation; tax allowances, tax deductible expenditure
Legal and regulatory	Patents, environmental regulations, health regulations, inspectorates, protection of designs, arbitration services, monopoly regulations, planning permissions for buildings and enterprises
Educational	General education, universities, technical education, apprenticeship schemes, continuing and further education
Procurement	Defence purchases, central government purchases and contracts, local government purchases, R & D contracts, prototype purchases
Information	Information networks and centres, libraries, radio and television, freedom of information, advisory services, statistical services, government publications, data bases, museums, exhibitions, liaison services
Public enterprise	Innovation by publicly owned industries, setting up of new industries, pioneering use of new techniques by public corporations, correction of imbalances by public enterprise, participation in private enterprise, investment by public corporations
Public services	Investment and innovation in health services, public building, civil engineering and construction, transport, telecommunications, consumer protection
Political	'Atmosphere', honours system, intervention vs. non-intervention, regional policies, labour policies
Scientific and technical	Technical standards, government research laboratories, testing stations, support for research associations, learned societies, professional associations, research grants
Commercial	Trade agreements, tariffs, currency regulations

Source: After Braun, E., 'Government policies for the stimulation of technological innovation', Working Paper WP–80–10, IIASA, Laxenburg, January 1980.

innovative activity rather than at specific innovations; (iv) specific innovations, i.e. support for a particular new technology. Table 5.2 gives a few examples which show how different types of policy measures can be targeted.

Policy measures can act in any phase of the innovation process described in Chapter 2. The innovation process may be viewed as a kind of electrical network in which all the switches have to be appropriately set before the innovation can proceed from one phase to the next. The setting of switches, symbolic of the actions required by human actors, also visualises the points at which policies can aid the required actions.[3]

The zeroth phase of innovation, in which consideration is given to the nature of the firm and the world it lives in, is strongly influenced by the ambience of free or restricted trade, availability of credits, availability of research facilities, information networks, monopolies and other tangible facts of political and economic life. But intangibles of ambience also play a role—personal honours and social hierarchies, the aura of success, public opinion.

The first phase of the innovation process—the emergence of an idea for innovation consisting of the confluence of a new technical possibility with a market need—requires easily accessible information systems to provide both market and technical intelligence. Many, though by no means all, ideas for innovation emerge from research laboratories and here support from public funds can be of the essence. For the research stage of an innovation can only be financed either from internal or from public funds; normal loan finance, however risk orientated, will shy away from the incalculable risks of research. The support of pure and applied research has therefore traditionally been a cornerstone of public support policies for technological innovation. This support can be in the form of direct or indirect research subsidies, in the form of publicly financed laboratories or publicly aided research associations. Sometimes the best support is in the form of liaison and information, for innovative ideas generally spring from the fusion of facts and ideas from previously separated areas.

To get from the first phase of an innovation to the second—the development phase—requires aid which is somewhat similar to that of the first phase, for more research and development work needs to be done and the product may still not be tangible enough for commercial venture financiers to be very interested. The innovator must have good knowledge of available materials, techniques, specialist suppliers, designers and, last but not least, advisers on patents and licences. This means that the innovator needs to be

Table 5.2. Examples of policy measures and their targets

Type of measure	Target			
	Ambience	Industry	General innovation	Specific innovation
Financial	Ease and cost of credit	Investment in regional factory building	Making venture capital readily available	Supporting specific R & D programmes
Taxation	Supporting entrepreneurial spirit	Making investment allowances	Allowances for innovative investments and R & D expenditure	
Legal and regulatory	Patent laws; monopoly regulations	Factory legislation	Health and safety regulations	
Educational	General educational provision; support for higher education	Technical training schemes		Training schemes in specific new areas
Procurement	Level and type of public expenditure	'Buy at home' policies	Procurement specifications	R & D contracts and orders for specific new equipment

Information	Libraries, broadcasting, government statistics	Technical information services	Liaison services	Information programmes on specific new technologies
Public enterprise	Strength of public sector	Active regional policies	Participation in new ventures, innovative policy in state industry	Public enterprise in new technology; specific innovations in public enterprises
Public services	Transport and communications			Development and use of specific innovations
Political	Access to information; Public opinion			
Scientific and technical	Technical standards		R & D availability from public sources	Specific R & D support
Commercial		Tariffs, trade missions		International cooperation in new ventures

Source: Braun, E., 'Government policies for the stimulation of technological innovation', Working Paper, WP-80-10, IIASA, Laxenburg, January 1980.

a member of an efficient information network. As at this stage much effort and money are consumed, financial assistance can be crucial.

In the past most government innovation support measures stopped with the end of the second phase, when a finished prototype emerged. Marketing and starting up production of even the most revolutionary products were regarded as purely commercial activities, for which public support was both unnecessary and repugnant. More recently, thinking has changed a little and some support programmes extend right up to the marketing stage. As the third innovation phase, the implementation stage, is the most expensive and also exposes the greatest weaknesses in the scientific/technical first-time entrepreneur, such extensions of support must be regarded as eminently sensible. Policy measures should not only be designed to stimulate innovation, they should also aim to remove difficulties at specific points of the innovation process. Hence an understanding of the process—a theory of innovation—is an integral part of innovation policy.

The above classification system of policy measures for the support of innovations is not unique. Different classifications have been used by different authors. An OECD report of 1978 lists three basic types of policies for the stimulation of innovation.[4]

(a) Specific measures:
  (i) instruments aimed at establishing and/or maintaining an interface. Examples are:
  '—instruments for promoting research associations
  —brokerage instruments either combined with or without technical or financial aid, etc.'[5]
  (ii) instruments based on loan or subsidy procedures, providing direct financial aid to innovation.
(b) Non-specific measures:
  (i) measures connected with science policy, especially involvement of government laboratories both in setting standards and directly aiding innovation;
  (ii) measures combined with instruments of industrial policy;
  (iii) measures grafted onto regulations for controlling relationships between various economic transactors; these measures include patents, licensing, consumer protection, procurement, taxation, etc.
(c) The final class of measure is called 'complex measure' or 'major programmes' and these are often 'leading instruments of sectoral policy'.[6]

The then German Federal Minister of Research and Technology,

V. Hauff, simply classified instruments according to the market at which the innovation was aimed.[7] In this view, a set of stimulation policies can aim at innovations for the private purchaser, the industrial purchaser, or the public purchaser. The first set of measures will essentially aim at new or improved consumer goods, the second at capital or investment goods and the third will aim to arrange procurement in such a way as to stimulate innovative activity.

Yet another alternative classification considers the mechanisms by which governments can stimulate innovation.[8] The authors list twelve such mechanisms and these fall into three categories: initiating mechanisms, sustaining mechanisms and restructuring mechanisms. Into the first category fall activities such as 'reducing the probability of technical or commercial failure'. In the second category we find measures such as 'increasing transfer of technical knowledge between institutions' and in the third category are measures 'influencing labour's receptivity to technological change and internalising the human costs associated with innovation activity'.

So far we have considered only government support for technological innovation, but other purposes of technology policy are equally possible. Government may wish to assure maximum safety at work, or the cleanest possible air, or the least consumption of scarce raw materials or any other conceivable objective of technology policy. In each case different sets of possible measures with different target areas can be identified. Let us pick just one more example and assume that a government wishes to ensure minimum consumption of scarce raw materials. The easiest way to achieve this is, of course, to reduce all technical and production activities to a bare minimum. This, however, would be incompatible with other government objectives, such as the maximum possible standard of living for the population, greatest possible influence in world affairs, free trade, and what we have called the new mercantilism. The compatibility or incompatibility of different objectives is one of the many difficult problems of policy making.

In order to find a set of possible policy measures for our assumed objective of minimum consumption of scarce raw materials, we may view the manufacturing–consumption cycle—in principle and with total disregard of all detail—as shown in Figure 5.1. It is apparent that the consumption of raw materials depends on total manufacturing activity, total consumption, design and life-time of products, degree of recycling of used materials, and efficiency of extraction of materials from their raw source. Policy measures can attack all these aspects. For the sake of brevity the types of policy measures which can be used for the achievement of minimum

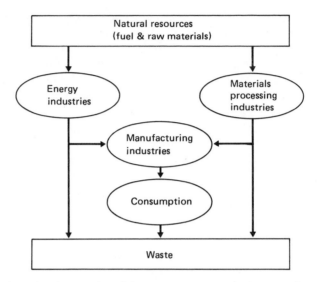

Fig. 5.1. Schematic representation of the resource–consumption–waste cycle.

raw materials consumption and their target areas are tabulated in Table 5.3, in analogy with Table 5.2, which shows measures for the stimulation of innovation.

ACTUAL POLICY MEASURES

The discussion and examples so far have tried to give an idea of the possibilities available to government, in a capitalist or mixed economy, to influence the path technology takes. The discussion was anything but exhaustive, and only intended to indicate the room for manoeuvre in which government can operate. A few examples of actual government measures of recent years in different countries will serve to illustrate how governments use the scope they have.

Among the small sample of policy measures purely infrastructural and ambience measures will be omitted. The former because it is obvious that all developed countries have an infrastructure of roads, railways, telecommunications, patent laws, universities, schools, apprentice schemes, technical standards, health and safety regulations, government laboratories, and all the rest of the paraphernalia which are indispensible for a modern industrial state. Ambience measures, on the other hand, are too subtle to describe in a general way. What constitutes the ambience of a particular country needs extensive detailed knowledge of many tangible and many more intangible factors; much can be described but even more resides in hidden assumptions which the uninitiated cannot begin to comprehend. All the programmes described below are aimed at the stimulation of innovation; a major preoccupation of most governments in recent years.

An example of a financial support programme aimed at general innovative activity, perhaps with special emphasis on small and medium sized enterprises is the *Erstinnovationsprogramm* of the German Federal Ministry for Economic Affairs.[9] In this programme the Ministry pays 50 per cent of R & D and pre-production costs for 'important programmes of high risk, which would not be launched without aid or would progress much more slowly'. If the project is commercially successful, the grant has to be repaid within ten years.

In a somewhat similar scheme by the German Federal Ministry for Research and Technology, aimed purely at research and development contracts placed by small firms with external organisations, the Ministry pays 30 per cent of the cost of up to DM120,000 per enterprise, per year.[10] The measure is to promote 'R & D contracts which aim at obtaining technically new or improved products or processes and which contribute to an improvement in the economic performance of the enterprise'. The scheme is a powerful instrument of public policy in that it not only supports innovation, but specifically excludes R & D contracts 'which run counter to the public interest'. The phrase 'counter to the public interest' does, of course, pose the question of determining what the public interest is. However, without the hidden assumption that government is the guardian of public interest and therefore can find ways and means of determining this interest in each case, all public policy-making in the current sense collapses. Thus the true essence of technology policy, the question of what interests it should and can serve, is hidden between the lines of policy pronouncements, in phrases which sound as innocent as they are impotent.

The theoretical foundation for recent British research policy, and thereby for much innovation policy, was laid in a White Paper, *Framework for Government Research and Development*, published in 1972 and based on previously published reports by Lord Rothschild and by the Council for Scientific Policy.[11] The famous customer–contractor principle was laid down which stipulates that research shall be done by a 'contractor' in response to a requirement formulated and financed by a 'customer'. 'Departments, as customers, define their requirements; contractors advise on the feasibility of meeting them and undertake the work . . .'.

This is music to the ears of every market-orientated economist,

Table 5.3. A few possible ideas for measures to implement a materials conservation policy

Type of measure	Target area				
	Ambience	Industry	Commerce	Foreign trade	Consumers
Financial	Cheap long-term loans, expensive short-term money	Support for materials saving design; support for recycling; research grants for materials conservation	Incentives for collection of used materials	Tariffs on selected materials	Incentives for recycling; support for repairs to make goods last longer; long-term loans for consumer durables
Taxation	Lowering of relative cost of labour vs. capital	High tax on new materials; lower rates on recycled materials; similarly, high tax on new machinery, low on reconditioned	Tax incentives on recycling operations and material collecting services		Decreasing tax rates for older durables, including cars
Legal and regulatory		Control of waste materials; controls on writing off of machinery		Import restrictions on selected materials	
Educational	'Old is beautiful' education	Materials engineering courses			

Procurement		Minimum arms purchases; specifications written to save materials			
Information	'Waste not, want not' campaign	Recycling and materials saving information centres			Propaganda on value of materials; 'old and working well is as good as new'
Public enterprise		Research on re-cycling, on materials substitution and on materials saving design	Public trading companies for materials		Nationwide repair services; material collection services
Political	Peace in our time—for war is the greatest waste of materials			Re-negotiation of trade agreements	
Scientific and technical		Research institutes for materials conservation; support services for materials conservation and substitution			

but in practice the principle holds only if the concepts of 'customer', 'requirement' and 'contractor' are stretched well beyond their common meanings. In the largest publicly financed slice of R & D, within the remit of the Ministry of Defence, the customer–contractor principle has always approximately applied, even before it was enunciated. Civil R & D, on the other hand, is so diverse, and relies so heavily on a wide variety of sources of ideas, that any real application of the customer–contractor principle could only act to stifle and inhibit creativity and inventiveness.

Many departments of government now have their own chief scientist who organises the work of Research Requirement Boards. The Department of Industry, one of the chief agents of innovation policy, has five such boards, each covering a broad area of industrial activity. The boards act as the departmental formulators of research requirements and these are often broadly interpreted to mean innovation support. Industrial and other interests are strongly represented on the boards and they do form strong consultative links, so that they formulate R & D requirements in conjunction with industry and research laboratories. The collusion and confusion between contractor and customer, so necessary to the proper functioning of the system, is thus assured.

The boards prefer, in principle, to fund projects in direct support of the government's industrial strategy, i.e. they give preferential treatment to projects in support of technologies singled out as of high priority. The support is not, however, exclusive and anybody with a good project can apply and has a chance of being accepted. Generally, projects should 'yield financial returns and improve the manufacturing performance in terms of international competitiveness, added value per employee and energy conservation'. What more could anybody ask?

The working of these arrangements has recently been reviewed and apparently found satisfactory.[12] Quite large sums are dispensed in this way. For example the Mechanical Engineering and Machine Tools Requirements Board had an annual budget of £10 million in 1979. But to put the total activities of the DoI in perspective, it must be realised that it controls only about £300 million of R & D expenditure and this represents about one-seventh of total private industrial R & D in Britain.

Although the apparatus described so far is designed to support R & D, it has recently become increasingly clear that major obstacles to innovation occur in late phases of the process and some support has been shifted into these areas. Several schemes are designed to go much beyond the R & D stage, as for example the DoI's Product and Process Development scheme, first announced in July 1977, which provides up to 50 per cent government funding from the design stage up to the point of commercial production.[13] The scheme allows either for grants of up to 25 per cent of qualifying costs or for a 50 per cent contribution coupled with a levy on sales. Furthermore, under certain conditions the DoI, under its Pre-production Orders scheme, is prepared to purchase pre-production models of new types of industrial equipment, normally costing between £25,000 and £1.25 million, and lend such equipment for evaluation to potential customers. If the customer is satisfied, he is expected to buy the equipment at the end of the one or two year trial period. If not, the equipment manufacturer is expected to buy it back.

The R & D support dispensed under the general schemes supervised by the Research Requirement Boards is similar in kind to that distributed under special schemes announced by the DoI from time to time. Currently the special schemes deal mainly with the fashionable technologies: microelectronics and information technology on the one hand and biotechnology on the other.

The basic strategy of all support schemes attempts to stimulate the availability of general advice and information; special consultancy advice; grants for research and development; and aid for product launch or the introduction of new production machinery, as applicable. We shall describe only the Microelectronics Application Project (MAP) in some detail.

In mid-1978 the British government became extremely concerned about the lack of awareness among British managers of the opportunities and challenges offered by microelectronics. A survey conducted at the time found that only about half the firms in manufacturing industry were aware of the existence of microelectronics and that only 5 per cent of firms were actually doing something about it. The Department of Industry then initiated several new support programmes for microelectronics under the Microprocessor Application Project (MAP) and one important aspect was 'to alert managements to the industrial scope and potential of microprocessors and to assist in retraining staff'.[14] Under this scheme numerous conferences were organised, supported by DoI funds and publicity material. Training courses in colleges and universities also received support under the scheme. It is thought that the awareness and training programmes, combined with substantial media coverage and the publication of numerous books, did raise the awareness of microelectronics to a pretty high level over a period of about two years.

Other government departments are making their contributions to the training and education programmes. The Department of Education, in conjunction with the DoI, is supplying all schools with small computers (though difficulties seem to arise from a shortage of cash to buy software and books). The Department of Employment, through its Manpower Services Commission is supporting many training courses for young people and retraining courses for unemployed older workers. The latest development in this activity are 6 to 12 months courses for young unemployed people, run in conjunction with so-called Technology Centres. These are advisory centres on the use of microcomputers and are administered by the National Computing Centre. The Department of Industry is providing initial financial support for the centres, but within two years they will have to earn enough income from consultancy and training services to become self-financed.

Traditionally, the manufacture of electronic components in Britain has been supported in great measure by the Ministry of Defence (MoD). The MoD has paid and still pays for much advanced research and development work and also is a customer of some substance for electronic components and devices. Plans are rumoured for a huge investment of £50 million into the development of very high speed integrated circuits; a path the US military have also been treading.

The MoD is by no means alone in its support for the electronic component industry. Even in early 1978 DoI support for the industry was running at £20 million per annum—40 per cent of this for semiconductors. In 1978 a scheme called Microelectronics Industry Support Programme (MISP) was announced.[15] This scheme was supplementary to MAP and pledged £70 million for a five-year programme aimed at helping companies to establish the manufacture of both special purpose and standard microelectronic components. MISP also aimed to support potential supplies of manufacturing equipment for the microelectronics industry and perhaps radically new alternative processes which might prove important for the industry.

The picture would be very incomplete indeed without mention of INMOS. This is a company founded in 1978 and financed to the tune of £50 million by the National Enterprise Board (now British Technology Group or BTG). The purpose of the company is to manufacture VLSI (very large scale integration) chips only, so as to become specialists in the high technology end of this high technology industry. The first products announced by the company were a 16K static RAM and a 64K dynamic RAM. These are probably to be followed by very advanced microprocessors, perhaps single chip micro-computers.

Although BTG money essentially comes from the Treasury, the enterprise was set up by three entrepreneurs and two of these are American. In fact the company has manufacturing and design facilities in both the United States and Britain and, arguably, this dual base and with it access to American know-how, is a necessary condition to make eventual success possible.

Many observers believe—and some official reports support this belief—that for most countries it is more important to apply micro-electronics than to manufacture integrated circuits and other micro-electronic components. Outstanding among the applications is, of course, the manufacture of electronic equipment of all kinds: telephone exchanges and other telecommunications equipment, computers large and small, instruments of all kinds, process control equipment, electronic office machinery, electronic games and all the rest of the known and still to be invented electronic gadgetry. Many of the items listed above are, in a loose sense, free standing and self contained items of electronics. No less important are applications of microelectronics in conjunction with mechanical machinery. To list but a few: computer numerically controlled machine tools, robots, computer controls for automobile engines, electronic weighing machines and computer controlled warehouses.

It is argued that the addition of modern electronics to many traditional machines can breathe new life into them and thus create new markets on the one hand and ensure competitiveness on the other. These considerations are reflected in several policy measures. Thus the previously mentioned Microprocessor Application Project (MAP) has, in addition to its information function, two strands in support of microelectronics applications. One consists of help with feasibility studies for new products, normally carried out by consultants registered by the DoI for this purpose. By late 1980 nearly 800 such studies had been completed and 75 per cent of their projects were being further pursued. The second strand consists of financial help to firms developing products using microelectronics. By late 1980, 345 such developments had been supported with grants of up to 25 per cent of qualifying costs or an alternative cost-sharing arrangement. Currently, a 'special offer' of $33\frac{1}{3}$ per cent grants is available.

More recently, both the British Technology Group and the Department of Industry have re-stated their faith in microelectronics application and have expressed their willingness to back their faith with money. BTG is giving high priority to firms wishing to develop and market innovations using, or related to, microelectronics and information and the DoI is planning to spend £80 million over the next four years on the support of information technology, under their Product and Process Development Scheme.

The Product and Process Development Scheme straddles the very narrow boundary between the development and diffusion of new technology; that is between the development and implementation phases of technological innovation. While the DoI will encourage the applications of certain new manufacturing technologies, e.g. Computer Aided Design (CAD), Computer Aided Manufacture (CAM), and robots, it will also assist firms wishing to develop these technologies. In fact the interaction between supplier and user of advanced techniques is a vital one and many a project has failed because of inadequacies in this relationship. The Product and Process Development Scheme recognises these facts in assisting user trials through financing pre-production orders for new equipment.

Most recently robotics has been recognised as a priority area for government assistance and this spans the whole wide range of activities from fundamental research to development of robotic technology to the support of users of robotics. It must be stressed here that government is by no means the main agent of change. Robot manufacturers have perfected their products to a viable stage and have developed an adequate infrastructure of support for their potential clients. An association of industrialists and researchers, the British Robot Association, also plays an important promotional role.[16] Similarly, British industrialists have clubbed together to form the United Kingdom Information Technology Organisation and one of their specific aims is to influence government policy on information technology by presenting an industrial viewpoint.

Whether as a result of this lobby or otherwise, the government has become very active in the promotion of information technology, including the designation of 1982 as Information Technology Year. One of the ministers in the DoI has responsibility for this area and the IT Year is coordinated by a committee with industrial membership and an industrial chairman. A great deal of money and publicity were lavished on IT Year, which coincided with far-reaching decisions on the introduction of cable television in many British cities, thus laying the foundations for a vast flood of information, or what passes for it, into every British home and office. Among the many activities of IT Year, DoI has created an electronic show office, where all the latest advanced office equipment is demonstrated and advice given to potential users.

No attempt has been made here to give a fully comprehensive

description of all government activity in the field of IT and micro-electronics, but even from what has been said the reader will have gathered three strong impressions:

(i) the British Government is trying very hard and spending a lot of money to help British industry enter the microelectronic age both by using and producing the new equipment;

(ii) while some uses of microelectronics are relatively straight-forward and do not require government intervention, other areas are so fraught with difficulty and risk that even a non-interventionist government feels obliged to intervene;

(iii) the possibilities for uses of microelectronics—whether in products or processes or as ready-made equipment—are wide and varied, while the possibilities of entry into the manu-facture of integrated circuits are very limited indeed.

If the old liberal ideal, prescribing to the state the role of night-watchman, guarding the citizens unobtrusively against evil but not interfering with their legitimate activities, were still to hold, then it would appear that the night-watchmen have become engaged in quiet warfare among their own ranks. For modern governments vie with each other in the support of their industries and inter-national trading competition has partly been superseded by inter-national competition in support policies. Microelectronics and information technology support programmes are a prime example of this trend and very few governments are willing to be seen with-out such a programme. It would be monotonous to the extreme to describe them all, however sketchily, and we shall briefly men-tion only one more scheme, that of the Netherlands.

The government of the Netherlands, a small country which happens to be the home of one of the world's major electrical and electronic multinationals, promulgated its policy for microelectronics in 1981.

The government planned to:
— stimulate the improvement of working conditions and the training and re-training of employees;
— stimulate the application of micro-electronics in industry by means of subsidies, loans and advice;
— set up three micro-electronic centres for advice and consultation in service of small and medium sized firms; and
— make further technology assessment studies.[17]

The microelectronics centres were set up in 1982 and special credit facilities for firms wishing to introduce microelectronics have been made available. Total planned expenditure is as follows:[18]

Work and employment $18 million
Education   ?
Direct support for industry $8.1 million
Microelectronics centres $6 million
Technology assessment $0.6 million

Not all innovation support is concerned with microelectronics or uses methods of offering grants and technical support services. The old-fashioned method of tax incentives is still operative in many countries and is targeted at the enhancement of innovation in general, rather than on any particular technology. So, for example, capital expenditure on plant, machinery and equipment used for R & D purposes is completely tax deductible in the United Kingdom. Similarly, revenue expenditure on R & D, such as wages, is tax deductible and costs of licence fees and patents may be written off against tax over a number of years.[19]

Grants and subsidies are also often used in general innovation support, not only for specific selected technologies. For example, in the German Federal Republic investments into research and development facilities receive a tax-free subsidy of 20 per cent of allowable costs up to DM500,000 and 7.5 per cent of higher costs. Small and medium-sized enterprises additionally receive a subsidy of up to 40 per cent of their expenditure on wages and salaries of research personnel, up to a ceiling of a wage bill of DM300,000 per annum.[20]

Another common target for innovation support measures is the increased use and utilisation of patents. Several such measures exist, for example, in the German Federal Republic.[21] The Patentstelle für die Deutsche Forschung provides a consulting service and loans for inventors who wish to patent their inventions and find industrial sponsors for further development and production. The Max Planck Gesellschaft, a major publicly financed research organisation, owns Garching Instrumente GmbH, which mediates the licensing of patents and develops and sells prototypes of equipment invented in MPG laboratories. There are two further patent information centres: the Arbeitsgruppe Patentverwertung is establishing a system to inform industry about patents resulting from government funded projects, while the Technologieverwertungsstelle der Grossforschungszentren aims at selling patents resulting from research at major publicly funded laboratories. The German Patent Office itself offers to produce, at a cost, surveys of patents existing in any specified area of technology even for companies not seeking to patent an invention. In the United Kingdom, one of the tasks of the National

Research and Development Corporation (now part of the British Technology Group) is the patenting and patent management of inventions originating in the public sector, particularly in government research laboratories and in universities.

A very large variety of other sources of technical information are supported from public funds. A United Kingdom Central Office of Information pamphlet[22] lists: general library services, including British Library, Science Reference Library, British Library Lending Division; and information services including those provided by the Department of Industry, the United Kingdom Atomic Energy Authority (e.g. Technology Reports Centre, Ceramics Centre, Non-Destructive Testing Centre), etc. The only item missing from the list is an information centre on information centres.

One of the widely quoted success stories of a mainly advisory and information system is that of a special working party set up by the German Federal Ministry for Research and Technology, together with the central 'applied science' research organisation (Fraunhofer Gesellschaft), the user industry, trade unions and the manufacturer's organisation to advise the clock and watch industry on technological change. Assistance has been given to firms in the watchmaking, clockmaking, sensor, weighing machine and office and sales equipment areas. Services available under this scheme include advice on new applications and diversification, technical advice and provision of know-how, preliminary assessments of development projects and advice on the availability of public funding for innovative activities.

The total funds available for innovation support are very large indeed. A recent estimate for the United Kingdom and the German Federal Republic is shown in Table 5.4. The number and variety of policy measures taken by various governments is also very large. A systematic attempt to list and discuss government support policies for innovation was made by Rothwell and Zegveld in their recent book on industrial innovation and public policy.[23] Particularly their chapters 4 to 7 deal with issues similar to those dealt with in our present chapter.

MISCELLANEOUS GOVERNMENT TECHNOLOGY POLICIES

Although the efficacy of support policies is very hard to determine, doubtless government intervention has had a considerable impact on some spheres of technology.

When it became clear that robotics was going to be an important field, particularly in its impact on the productivity and quality of

Table 5.4. Government funds available for innovation in current prices

	1972	1973	1974	1975	1976	1977	1978	1979
Germany F.R.								
	(DM million)							
Patentstelle für die Deutsche Forschung	0.5	0.6	0.7	0.9	1.1	1.2	1.5	
Technologieverwertungstellen der Grossforschungszentren							1.0*	
Deutsches Patentamt	95	101	104	117	123	125	131	
R.K.W.	30	32	34	40	41	42		
Fraunhofer Gesellschaft	62	68	82	90	108	126	138	
Technological Advisory Services to SMEs						0.5	2.6	3.0
VDI Technologie-Zentrum						0.7	0.7	1.0
R & D manpower grant programme								300
BMFT-Projektförderung					1,269	1,500*	1,500*	
Vertragsforschung							6.0	10.0
Wagnisfinanzierungsgesellschaft						0.8	1.8	5.5
Erstinnovationsprogramm	3.9	7.2	6.0	7.3	9.9	12.1	16.0	19.5

United Kingdom

(£ million)

NRDC (revolving fund)	4.46	3.84	2.49	3.17	4.14	4.39	6.48	7.03
Low Cost Automation Centres	0.10	0.10						
Industrial Liaison Services Centres	0.50	0.20						
Small Firm Information Centres	0.03	0.5	0.5	0.38	0.5	0.5	0.42	0.43
Research Associations	2.90	2.8	3.2	4.7	6.6	8.0	11.0	11.5
Small Firm Counselling Service					0.03	0.05	0.34	0.35
Manufacturing Advisory Service						0.23	1.86	1.64
Requirement Boards		0.30	0.70	1.80	2.00	2.20	10.00	11.00
Collaborative Development Contracts	4.50	8.42	10.39	10.14	10.34	11.13	14.86	16.57
Pre-production Order Assistance							0.38	
Product and Process Development Scheme							0.27	0.6
Software Products Scheme	0.03	0.06	0.11	0.31	0.27	0.20	0.40	0.6
Micro-electronics Support Scheme						0.21	6.0	6.0

*Approximate figures.

Source: Hagedoorn, J. and Prakke, F., *An Expanded Inventory of Public Measures for Stimulating Innovation in the European Community with Emphasis on Small and Medium Sized Firms*, report prepared for DG XIII of the Commission of the European Community, Brussels, 1979.

manufacture in the automobile industry, and when it became equally clear that German industry or research organisations were not doing much about it, the Federal German government decided to intervene. Undoubtedly this intervention was only one factor among many and perhaps the situation was ripe for something to happen. Whatever the causes, the facts are that since the announcement of the government policy German contributions to robotics research have soared[24] and German motor manufacturers are making extensive use of spot-welding robots of German manufacture.[25] Even Ford in Britain use German-made robots specially developed for Ford.[26]

A hotly debated topic of technology policy is the so-called humanisation of work. Nobody doubts that present working conditions in some industries are terrible and that particularly jobs on assembly lines are barely tolerable for humans.[27] The only way people put up with this kind of work is by taking the so-called instrumental attitude. Roughly speaking they say to themselves 'the job is lousy, but pay is not too bad and I need the money'.[28] After much political wrangling, the German Federal Ministry of Research and Technology, together with the Ministry of Labour and Social Affairs, announced in 1974 a programme whose aim was to support firms introducing new production machinery, provided the new machinery allowed better working conditions in both the physical and mental sense. The avowed overall aim is to make factory work more satisfying, more dignified, more human. The main components of the programme are: prevention of accidents at work; prevention of occupational diseases; removal of excessive work loads; improvement of the quality of work; reduction of negative effects of work upon other aspects of life; implementation research; dissemination of results of the programme. By mid-1978, about DM240 million had been spent. Of this sum, 53 per cent went on the reduction of excessive work-loads and 27.8 per cent on improvement in the quality of work.[29]

Obviously this kind of programme is a political hot potato and its true achievements are hard to assess. An attempt to evaluate the programme was made by several researchers and this is summarised in a recently published volume.[30] Whatever the direct results and whatever the possible abuses of this programme, it is one of the great pioneering efforts by an enlightened government to introduce deliberately humanitarian criteria into a situation in which previously only economic necessity counted and King Efficiency reigned supreme, except for occasional insubordination by his subjects, the workers.

Another hotly debated issue of the day is the role of government

regulations in innovation. As regulations concerning safety, environmental pollution, working conditions and descriptions of goods
became more ubiquitous, so their opponents began to argue that so
much money had to be spent by manufacturers on complying with
regulations, that none was left for innovation. Others argue the
opposite view and believe that the need to comply with new regulations gives a fillip to new products and processes. While dispassionate
writers like Lederman[31] think that no consensus exists on whether, on
the whole, regulations are beneficial or detrimental to innovation,
many members of a panel conducting a US Domestic Policy Review
of Innovation in 1977 were extremely critical.[32] Among the recommendations of the subcommittee on Industry Structure and Competition are that 'each regulatory agency should issue a long range
statement of regulatory intent . . .' and 'Where a company has related
compliance requirements controlled by more than one law, consultations must occur between the agencies.'[33] No doubt, some confusion
and changes of direction can cause difficulties. The main thrust of the
argument, however, is directed against regulations which prescribe
solutions instead of prescribing performance standards.

Innovation is negatively impacted by regulating the means rather than the
ends.[34]

Regulations should only prescribe the standards of performance to be achieved,
not the means by which they are to be achieved.[35]

Small firms, as so often, become the subject of special pleading:

Smaller firms, historically the source of significant contributions to innovation,
suffer disproportionately greater injury from the overall costs of regulations
than do larger firms.[36]

The sub-committee on Federal Procurement Policy makes similar
points in relation to procurement policy. The recommendations
of this committee on general policy include the following:

Express needs and program objectives in mission terms and not equipment
terms to encourage innovation . . .

. . . allow competitive exploration of alternative system design concepts . . .

Different sub-committees of the same Domestic Policy Review of
Innovation view obstacles to innovation and the role of the state
very differently. The Labor Advisory Committee states: 'The best
stimulus to innovation comes from a healthy full employment economy—not from weakening protections'. The sub-commitee on
Public Interest:

We do not accept the widely held industry assumption that regulations impede innovation. . . . Moreover, job security is a fundamental aspect of innovation. Frightened workers, worried about whether another job will be available if a specific new idea is implemented, often oppose innovation. . . . From the public interest perspective, the rate of innovation is subservient to the question of the direction of innovation.

Even these brief abstracts show clearly what an intensely political matter beliefs about the effects of government policies are. In this respect at least, technology policy differs but little from other areas of public policy.

If there is one industry above all others in which regulations might be expected to affect innovative activity significantly, it must be the pharmaceutical industry. But even in this industry the case remains unproven. There is no doubt that the rate of introduction of new drugs declined steeply from about 1959 in the United States and from 1964 in the United Kingdom. The decline in the United States thus anticipated by several years a tightening up of regulations, which occurred in 1962. The decline in the United Kingdom does not correlate with any particular tightening of British regulations. It is nevertheless imaginable that the sharp decline in the introduction of new pharmaceuticals, particularly marked in the United States, was partly initiated by anticipated regulations and that any recovery in the rate of introduction may have been prevented by the actual introduction of stricter rules. Steward has shown, however, that if new drugs offering significant gains in therapeutic effect are considered separately, then the rate of introduction shows no significant change with time and certainly no correlation with regulation.[37] It could be argued that if regulations caused any decline in the innovative activity of the pharmaceutical industry, then this decline did not prove detrimental to the health of the population.[38]

A recent summary of British government policy for the computer industry by White provides an example of the close proximity of science, technology, industrial, trade and education policies.[39] Early scientific and engineering efforts on computing and computers in Britain were not far behind those in the United States. By the time digital electronic computers became a commercial reality, however, American companies led the field. This did not deter several British companies from an effort to enter this developing industrial branch and a widely dispersed private enterprise effort, partially funded by the National Research and Development Corporation, followed. The rewards were meagre and many a company decided not to throw good money after bad. The balance of trade in computers between the United States and the United

Kingdom became extremely unfavourable and many a voice was raised in Britain calling for more British effort in this 'leading edge' technology. Government became closely involved in attempts to get several industrial firms to collaborate in an effort to leap-frog into an ultra-advanced 'supercomputer'.[40] Leap-frogging in technology rarely, if ever, succeeds—most efforts carry the analogy to its ultimate and end ignominiously, cold and wet.

During the sixties it became increasingly evident that only very few British firms could survive in the highly competitive computer business and that government assistance was becoming a condition for survival. Eventually a single company, ICL, was formed by merger in 1968, with government acting as a combination of mid-wife, provider of a dowry and prime customer by introducing a 'buy British' policy for all central government purchases of large computers. This policy has recently been rescinded, but government support for the struggling computer firm continues.

ACADEMIC CONTRIBUTIONS TO INNOVATION

Computing provides one of the happiest examples of truly important contributions of academic science to engineering developments and good feedback from industry into the university system. From the earliest days of computing universities played an important role and the very first British programmable electronic computer, EDSAC, was built at Cambridge University during the late forties.[41] In later years there was important collaboration between the University of Manchester and Ferranti Ltd, which led to the design and construction of Atlas, a large British computer of the early sixties. The most important British computer series of the seventies, the ICL 2900 announced in 1974, also was the fruit of collaboration between ICL and Manchester University.[42]

There are several good reasons for the relatively fruitful collaboration of the computer industry with universities, especially during the early immature years of the industry. In the early days there were very few experts available and certainly no steady supply of trained computer scientists. Such experts as there were had to be largely self-trained and opportunities for such spontaneous change of direction and creation of knowledge are greatest in the relative freedom of academic research. The computer industry did not manufacture large quantities of hardware. Its essential activity consisted of putting together theoretical knowledge and electronic hardware produced elsewhere. Thus universities could serve as a major supplier to the industry, a supplier of theoretical knowledge, while the electronics industry provided the hardware.

As computer science matured, the universities introduced proper courses and industry became assured of a supply of trained graduates. The growing computer firms could afford to put more resources into research until they were able not only to match universities but in fact to outstrip them by large margins. There still is some useful cooperation in research, as some talent still resides in universities, but by now the traffic of knowledge certainly operates both ways. Much knowledge is created in industry and the universities act as guardians, keepers and disseminators of this knowledge as well as that created within their own walls. Strong links between universities and industry are important, because they enable industry to gain access to the general pool of knowledge available in universities and universities to gather knowledge created in industry. In fact without feedback of knowledge from industry into universities the latter would become sterile purveyors of obsolete information. Thus an ideal relationship can be forged in which both partners gain more then they give.[43] In fact ICL have recently begun a new scheme to foster 'collaboration with British research institutions' under which they will spend a proportion of their research budget extramurally.

The contributions which university departments can make to technological innovation have probably been greatly exaggerated and thus expectations have been raised too high. Yet useful contributions do emerge and a useful symbiotic relationship does exist and needs to be fostered. In recent years many governments have made strenuous efforts to strengthen the links between the academic and industrial worlds in an attempt to make university research more cost-effective in the sense of producing more industrial benefits for the costs incurred by the public purse. In Britain, the Science and Engineering Research Council (SERC) have a whole range of programmes aimed at achieving 'relevance' of university research, where 'relevance' seems to mean a contribution to technological innovation and industrial performance. Although these schemes have had a degree of success and have helped to forge some useful and fruitful links between universities and industry, it must be hoped that the main purposes of universities will not be forgotten in attempts to increase their direct utility to the national economy. The main aims must remain the catering for intellectual needs and development of talents of the individual, the supply of highly trained people into all walks of life and the creation of knowledge unfettered by the urgent daily requirements of an industrial environment.

The primary function of universities, even in their role in support of industrial activities, is to supply industry and the industrial infrastructure with well trained people, equipped with up-to-date knowledge and, infinitely more important, the ability to acquire and use new knowledge.

Because of the large sums of money spent on higher education, many governemnts feel the need to try and increase the contributions made by university research to technological innovation. The feeling is abroad that universities are alive with numerous potential innovations and all that needs to be done is tap this potential for innovations to spring forth. Reality is a little more complex. From our description of the innovation process in Chapter 2, it would seem that the most important contribution universities can make (apart from the supply of highly trained personnel) is in their role as custodians of and suppliers to the 'common pool of knowledge'. For it is this general background of knowledge and techniques, obtained and maintained in painstaking 'pure' research, which supplies the missing links necessary to turn innovative ideas into innovations or, sometimes, provide the new technical possibility from which an innovative idea may spring. There is no prescription on how to increase the chances of pure research proving beneficial to innovation, except perhaps the need for close personal contacts in an information network consisting of both academic and industrial researchers.

The other possibilities of academic work becoming useful for technological innovation are through what we have called the 'instrument bond' and the 'innovation bond' between science and technology. If we assume, as a reasonable first approximation, that university research represents science (even if done in an engineering department) and industrial research represents technology, then the bonds linking science and technology become also the bonds linking university and industrial research and thus provide vehicles for academic contributions to innovation.

The instrument bond requires strong information links between scientists and instrument manufacturers. This link must consist of more than the mere distribution of sales literature and involve personal acquaintance. If general entrepreneurial assistance, as for example in an innovation centre, is available, then scientific instruments may occasionally attract entrepreneurial scientists into new ventures.

The innovation bond is more likely to involve university scientists in a peripheral—though important—way. The central innovation activity is much more likely to be located very close to normal manufacturing and commercial activities. The most likely role of university scientists is that of consultant and specialist adviser— roles of very considerable significance. Frequently university researchers may perform useful subcontracted investigations or tests as part of a product development programme. Occasionally product ideas other than for scientific instruments may emerge from universities, but overall universities are not really particularly suitable breeding grounds for this type of activity. Even when product ideas do emerge, they will generally require scrutiny by technical and commercial experts with wide industrial experience, and considerable industrial help with developing the idea into a product.

All the above links between the university research system and technological innovation can be strengthened by deliberate policy measures. There are numerous variants of these, but the basic idea is generally to establish, usually with the aid of some initial public funding, what are variously called 'science parks' or 'innovation centres' or variations on the theme.[44] The models for these ideas appear to be the entrepreneurial activity round MIT, on the famous Route 128, and the mistaken notion that silicon valley somehow owes its existence to the science park of Stanford University.

Innovation centres may go beyond merely catering for university-industry links and may play any or all of the following roles:

(i) the screening of ideas emerging from private inventors or research institutions and help with developing, patenting, market research, venture capital, public funding, selling licences or starting up production of new products;

(ii) keeping in touch with data banks of new products and helping small or new firms with starting up the manufacture of such products; help with obtaining licences, funding, market research and further development;

(iii) port of first call for manufacturers who do not know where advice might be available on new processes or equipment which may help them improve their manufacturing efficiency;

(iv) extension of above ideas to the service sector for the provision of new or improved services;

(v) provision of suitable premises for starting up high technology firms, including the sharing of some facilities.

The key element in the notion of a science park is that research carried out in the laboratories of universities or other institutions of higher learning can form the basis for new commercial activities in new or existing firms, and/or that universities can provide valuable advice and stimulation to a firm engaged in innovative activities. The inventor, corporate or private, needs to be put in touch with people who can competently and confidentially assess the technical feasibility of the idea, assess likely production problems and estimate probable markets. If an idea emerges unscathed from this scrutiny it becomes necessary to start developing it for production and at the same time seek patent protection on the one hand and financial backing on the other. The centre must be able to provide help with all these, as the inventor will not normally know where to turn. Thus the centre has to have close links with development laboratories, patent agents, financiers and government departments able to provide help.

The next stage of the innovation involves either selling licences to established producers or seeking a combination of suppliers and own production facilities. At the same time as preparation for production begins, active marketing must also get under way. Again, the centre will have to put the inventor in touch with suitable advice.

Finally, when it is decided to set up a company and start up production, the centre may be able to help with premises, management advice, advice on suppliers and markets and some shared services, e.g. workshops, design facilities, computer terminals, reception and telephone answering.

Although there are numerous institutions for the encouragement of innovation of vaguely this type in Germany, USA, the Netherlands, Ireland and Britain, only a few will be described briefly here.

The VDI Technologiezentrum (the technology centre of the association of German engineers) in Berlin

This centre is financially assisted by the Federal Ministry for Research and Technology (BMFT) and operates in four areas: (i) dissemination of technological information, especially bringing together potential suppliers with potential users of technology; (ii) information and advice on available federal aid and other finance; (iii) the centre owns and makes available some technical installations, e.g. microprocessor development systems; (iv) technological and marketing consultancy.

Innovation Centres in the United States

The first three Innovation Centres in the United States (separate and distinct from the Science Parks) were founded in the mid-seventies and had NSF funding of $3 million for five years. These centres are associated with the engineering and business schools at MIT (Massachusetts), Carnegie-Mellon (Pittsburgh) and the University of Oregon. The role of such centres is to help students and others to evaluate technology and R & D results, develop new products and services, provide assistance to independent inventors, establish new business ventures.

The emphasis in the different centres differs slightly. So at MIT the centre is to the inventor and innovator what a teaching hospital is to a medical student. At Carnegie-Mellon the main aid is with provision of 'seed money' and a shoulder to lean on. At Oregon the main thrust lies in the screening and evaluation of ideas. Several further centres have been founded since the mid-seventies.

The Innovation Centre for Small Industry in Limerick, Ireland

This is a venture run by several government and semi-government agencies. It has a modern purpose-built centre near the campus of the National Institute for Higher Education. The centre combines the roles of providing advice, consultancy, access to data banks and contacts with experts with that of providing workshop facilities and 'nursery' factory premises. The centre is adjacent to the Microelectronic Application Centre. It had an operational budget of £100,000 for the first year, but aims ultimately to become self-financing.

Science parks

The main feature of pure science parks, as opposed to innovation centres, is to make available suitable premises in close proximity to an academic institution. It is hoped to attract high technology entrepreneurs both from within the institutions of higher learning and from without and to develop close liaison between the businesses and the academic institution.

The Science Park in Cambridge, for example, is essentially a real estate development by Trinity College. Small industrial units were built on a landscaped plot of land owned by the college. The first tenant was a high technology entrepreneurial firm which emerged from the Cambridge University Engineering School in 1972. Since then a variety of outside firms has slowly been accrued, presumably attracted by the high prestige area, good access, pleasant landscaping and small industrial units to let. There are no common facilities and any links with the University are purely informal. The park simply functions as a modest commercial venture nurtured on the high prestige of Cambridge and its University.

Currently there is a flurry of activity in the United Kingdom, mainly supported by county council initiatives, to establish an array of science parks throughout the country. The idea has taken hold as a measure for industrial rejuvenation and the creation of new employment. To think that science parks might stem the tide of unemployment seems sadly reminiscent of the old tale about the little boy holding up the dyke by plugging the breach with his thumb.

SCIENCE POLICY

Despite all the efforts to increase the direct contribution of academic research to technological innovation, the hallmark of university research is that it is largely open and free from the constraints of immediate goal orientation. This means that a large proportion of university research is of a fundamental nature, or at least at some remove from the immediate day-to-day necessities of industrial research. For a variety of reasons it is to be hoped that the bulk of university research will retain these features.

The support of pure science has been in the public domain for a long time. Motives are mixed and range from the hope of aiding technological innovation or solving burning health or environmental problems, to support for the pure cultural activity of acquiring knowledge. Thus government support for science ranges from medical and environmental research, through applied physics, engineering and chemistry, right up to cosmology or elementary particle physics.

The setting of priorities and the total allocations for science budgets are the subject of permanent fierce debate. This is hardly surprising in view of the mixed motivation for the support of science and in view of the very large total sums involved. The questions posed to policy makers permit no rational answer, for how much is it worth to learn more about the origin of the universe or of life? Even in applied science a mixture of criteria of scientific timeliness and promise and probable economic pay-off must be judged and competing claims must be given priorities within limited budgets. An elaborate system of advisory bodies serves to form judgements on these matters.[45]

Perhaps because of the close links between science and weapons on the one hand and science and technological innovation on the other, scientific achievement has become one of the measures for national prestige and prowess. The total sums of money spent by governments on the support of research and development are stupendous, as Table 5.5 and Figures 5.2 and 5.3 attempt to show.

Table 5.5. Resources devoted to R & D and size of country 1977 (countries are ranked in decreasing order of gross intramural R & D expenditure)

	Expenditures			Personnel		
	GERD (NSE + SSH) ($ US million)	GDP ($ US billion)	GERD as % GDP	R & D personnel (thousands FTE)	Total labour force (thousands)	R & D personnel as % total labour force
United States	44,788	1,873.7	2.4	574.4	99,534	5.8
Japan	14,375	744.4	1.9	564.9	54,520	10.4
Germany	11,083	517.1	2.1	319.3	26,074	12.2
France	6,754	382.7	1.8	222.1	22,614	9.8
United Kingdom (1978)	6,688	314.5	2.1	n.a.	26,328	n.a.
Netherlands	2,116	106.5	2.0	52.9	4,880	10.8
Canada	1,856	200.0	0.9	56.0	10,579	5.3

Italy	1,909	215.4	0.9	97.3	21,794	4.5
Sweden	1,500	80.4	1.9	36.3	4,174	8.7
Switzerland	1,391	60.7	2.3	41.1	2,935	14.0
Belgium	1,059	79.3	1.3	29.5	4,056	7.3
Australia (1976/77)	920	95.4	1.0	43.6	6,313	6.9
Austria	589	47.9	1.2	n.a.	3,038	n.a.
Yugoslavia	528	45.6	1.2	22.9	5,161	4.4
Norway	503	35.8	1.4	13.7	1,851	7.4
Denmark	443	46.4	1.0	13.8	2,579	5.4
Finland	322	31.5	1.0	14.3	2,284	6.2
New Zealand	125	15.2	0.8	8.2	1,230	6.7
Ireland	75	9.4	0.8	5.8	1,145	5.1
Portugal (1978)	57	17.8	0.3	6.5	4,177	1.6
Iceland	13	2.0	0.7	0.6	98	6.4

NSE: Natural Sciences and Engineering. SSH: Social Sciences and Humanities. FTE: Full-Time Equivalent.
Source: Science and Technology Policy for the 1980s, Paris, OECD, 1981, p. 18.

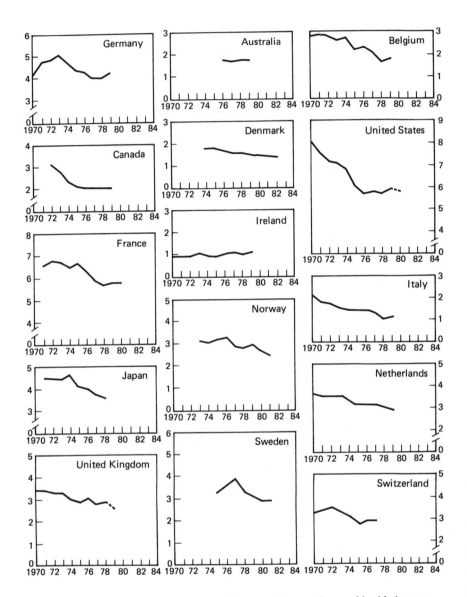

*As the definition and coverage of total public expenditures varies considerably between countries, comparisons between countries should not be made of levels but only of trends.

Fig. 5.2. Trend of government R & D expenditure in percentage of total government expenditure* in some OECD countries. *Source: Science and Technology Policy for the 1980s*, Paris, OECD, 1981, p. 17.

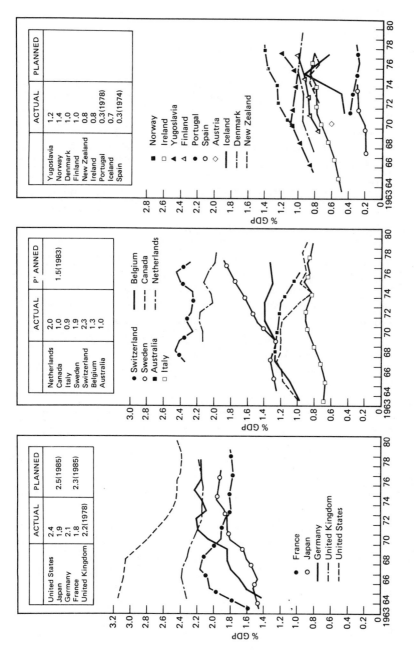

Fig. 5.3. Domestic R & D expenditure as percentage of Gross Domestic Product. *Source: Science and Technology Policy for the 1980s, OECD, Paris, 1981, p. 16.*

MILITARY R & D AND PUBLIC PROCUREMENT

The proportion or totals spent on defence R & D are by no means clear from the above figures. Because of problems of definition these expenditures are difficult to ascertain, but Pavitt and Soete[46] estimate that Britain spent 46.4 per cent of total government R & D expenditure in 1975 on defence, while the relevant figures for France and Germany (FR) were 29.6 per cent and 11.0 per cent, respectively. Kaldor[47] estimates that expenditure on British defence R & D amounted to £870 million in 1977. The White Paper *Review of the Framework for Government Research and Development*[48] suggests that in 1977-78 the Government spent £802 million on defence R & D while it spent £285.5 million on civil R & D. The discrepancies between the figures arise because of problems of definition and various omissions in different statistical compilations. It is clear, in any case, that expenditure on defence R & D is stupendous.

Many writers have argued that military R & D, as well as military production, severely distort civil technological innovation.[49] The arguments are pretty convincing, but would require a chapter of their own to be properly discussed. In particular the argument that military R & D deprives civil R & D of financial and manpower resources and the argument that easy profits on defence orders distort civil innovative effort deserve considerable attention.

Dickson has argued that in the British case at least, excessive concentration on military electronic projects led to a distortion of industrial effort which proved detrimental to commercial enterprise.[50] Several people in US electronic firms have argued similarly, though few would deny that military procurement helped the industry to overcome many a cash-flow problem.[51]

While it is impossible to overestimate the importance of military procurement in the US economy and generally in the scientific-technical system, civil procurement can also play a very major role. It has been said that US defence purchases give a vital fillip to advanced industrial technology, especially in electronics, computers and aircraft. It is equally often said that the Japanese economy thrives because it does not carry a heavy burden of defence expenditure. Certainly the Japanese example proves that electronics and computers can thrive without military orders.

A recent study by Müller and co-workers investigated the effect of capital investment by the German Post Office.[52] Only their summary is shown here in Figure 5.4 and Table 5.6. Through the action of various multiplier effects an investment of DM10 billion causes an overall expenditure of DM17 billion and gives employment

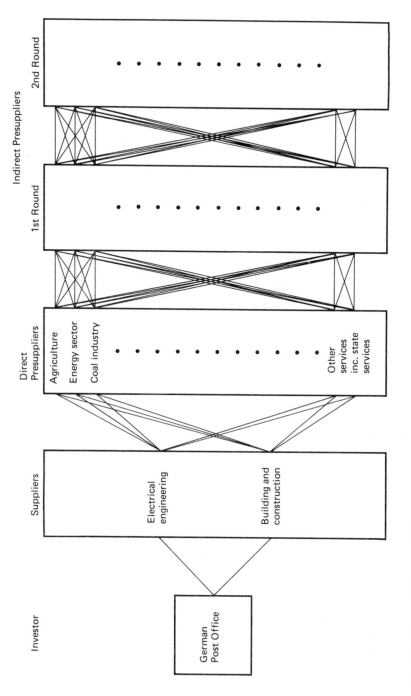

Fig. 5.4. Schematic diagram of direct and indirect production and employment effects of investment. *Source*: Müller, J. *et al*., *Direkte und indirekte sektorale Auswirkungen*, Deutsches Institut für Wirtschaftsforschung, Berlin, 1982.

Table 5.6. Employment effects of investment by the Post Office

	(DM thousand million)
(1) Investment in telecommunications in the FRG	10.0
(2) Production effects of investments in telecommunications in the FRG	
Production effects total 1980	17.9
Direct production effects	10.0
Supplies for direct production	6.8
Reinvestment effects and supplies generated by them	1.1
Total production effects of additional investment in telecommunications of DM1 thousand million (based on 1980)	2.5
	(Persons)
(3) Employment effects of investment in telecommunications in the FRG	
Total employment effect 1980	163,000
Direct employment effects	105,000
Employment effects of supplies for direct production	49,000
Employment effects of reinvestments	9,000
Total employment effect of additional investment in telecommunications of DM1 thousand million (based on 1980)	22,000
(4) Employment effects of DM1 thousand million investments in telecommunications (without reinvestment effects)	
In the FRG	15,100
In France	14,300
In Great Britain	38,500

Source: After Müller, J. *et al., Direkte und indirekte sektorale Auswirkungen* . . . , Berlin, Deutsches Institut für Wirtschaftsforschung, 1982, p. 5.

to some 163,000 people. Of these about 105,000 are employed by direct effects of the initial investment. Thus the expenditure of DM1 billion gives employment to 15,100 people. The equivalent sum spent in France employs an estimated 14,300 people while in Britain, because of the much lower labour productivity, an estimated 38,500 persons would find direct employment for the same investment.

EFFICACY OF POLICIES

If the formation of technology policy seems difficult, the problem of determination of the efficacy of policies is formidable. It is not too hard, though by no means trivial, to assess the direct effect of a control policy, say exhaust emission controls, as there are scientific methods able to measure air pollution before and after. It is more difficult to know whether the policy has made a positive contribution to the health and well-being of the population. To know the effect of an innovation support policy is harder still and a good deal of methodological effort will have to be made before truly useful feedback to the policy makers can be obtained. The German concept of *Begleitforschung*, where researchers are involved at all stages of implementation of a policy, is extremely useful in this context, but the paucity of knowledge on the efficacy of policies remains considerable. One of the OECD attempts to investigate these questions used the method of questionnaire addressed to government officials administering the various schemes identified by the report. The conclusions are somewhat confused, but on the whole officials either felt unable to answer certain questions or answered in a spirit of rosy self-satisfaction. To make matters worse, the questions were directed, in the main, at the attainment of intermediate, rather than final, objectives.[53]

Allen *et al.* investigated 164 innovation projects in several industries and several countries. They found that nearly half the projects were affected by government policies, but found no indication that success or failure were influenced. Projects involving a response to regulations tended, on the whole, to have an above average rate of success.[54]

Rubenstein *et al.* investigated management perception of the role of government incentives to technological innovation in Britain, France, West Germany and Japan. By and large, they found that managers tended to regard government incentives with great scepticism, although Japanese and German managers less so than their British and French counterparts.[55] Often, it would seem though scepticism is no barrier to acceptance of government money.

Jones and Willett discuss the difficulties in evaluating the costs and benefits of government laboratory research and information services. They do believe that these services make a positive contribution to innovation, but find a rigorous examination of all costs and benefits too complex to be cost effective.[56]

Thus the answer to the vital question on the efficacy and effects of government policies for technology remains elusive and poses

Table 5.7. Air pollution in London

Year	Maximum mean daily concentrations in Central London (mg/m^3)		Estimated extra deaths in Greater London
	Smoke	SO$_2$	
1952	6,000	3,500	4,000
1962	3,000	3,500	750
1972	200	1,200	Nil

Source: Royal Commission on Environmental Pollution, *Fourth Report*, London, HMSO, 1974, p. 14.

a major challenge to both researchers and policy makers. Some policies such as the British clean air acts, appear to achieve spectacular success, as shown in Table 5.7.

Unhappily, even this apparently brilliant success of a policy cannot be uniquely ascribed to government measures. For the substitution of gas for coal and the move away from traditional open coal fires was becoming fashionable anyway in the late fifties and early sixties for reasons of comfort, convenience and economy. Undoubtedly policy measures aided the transition both by regulation and by powerful financial incentives in favour of smokeless fuels.

The success or cost effectiveness of many measures remain a matter of conjecture. Worse still, many urgent actions await international agreement, while the environment suffers irreparable damage by acid rain, nitrous oxide, deforestation, excessive fishing and many other abuses.[57]

Perhaps one may be forgiven for sometimes wondering whether the efficacy of policy measures is inversely proportional to their ease of administration.

NOTES

1. Talk by Kenneth Boulding at International Institute of Applied Systems Analysis, Laxenburg, summer 1982.
2. Braun, E., 'Government policies for the stimulation of technological innovation', Working Paper WP-80-10, International Institute of Applied Systems Analysis, Laxenburg, 1980; Braun, E., 'Government policies for stimulating technological innovation', in Maier, H. and Robinson, J. (eds), *Innovation Policy and Company Strategy*, Laxenburg, International Institute for Applied Systems Analysis, 1982, pp. 51-66; Rothwell, R. and Zegveld, W., *Industrial Innovation and Public Policy*, London, Frances Pinter, 1981.

3. Braun, E., 'Constellations for manufacturing innovation', *Omega*, **9** (1981), pp. 247–53.
4. *Policies for the Stimulation of Industrial Innovation*, vol. 1, Paris, OECD, 1978.
5. Ibid., p. 34.
6. Ibid., pp. 35–6.
7. Hauff, V., in Karl A. Stroetman (ed.), *Innovation, Economic Change and Technology Policies*, Basle, Birkhauser Verlag, 1977, p. 347.
8. Allen, T. J., Utterback, J. M., Sirbu, M. A., Ashford, N. A. and Hollomon, J. H., 'Government influence on the process of innovation in Europe and Japan', *Research Policy*, **7** (1978), pp. 124–49.
9. Hagedoorn, J. and Prakke, F., *An Expanded Inventory of Public Measures for Stimulating Innovation in the European Community with Emphasis on Small and Medium Sized Firms*, reports prepared for DG XIII of the Commission of the European Community, Brussels, 1979.
10. *Förderfibel*, Bundesministerium für Forschung und Technologie, Bonn, 5th edition, 1982.
11. *Framework for Government Research and Development*, Cmnd 5046, London, HMSO, 1972.
12. *Review of the Framework for Government Research and Development*, Cmnd 7499, London, HMSO, 1979 and *Science and Technology Policy for the 1980's*, Paris, OECD, 1981, p. 27.
13. Department of Industry, *Product and Process Development Scheme*, London, 1977.
14. Department of Industry, *Microprocessor Application Project*, London, 1978.
15. Department of Industry, *Microelectronics Industry Support Programme*, London, 1978. See also a series of brochures currently published by the Department of Industry on various innovation support schemes.
16. Fleck, J., *The Diffusion of Robots in British Manufacturing Industry*, final report to the Leverhulme Trust Fund, 1983; Fleck, J., *The Introduction of Industrial Robots*, London, Frances Printer, forthcoming.
17. Loo, van der, H. and Slaa, P., 'Information Policy in the Netherlands', proceedings of FAST conference, *Information Technology and the Sharing of Power*, Copenhagen, November 1981.
18. Ibid.
19. Hagedoorn and Prakke, *An Expanded Inventory of Public Measures*.
20. *Förderfibel*, Bundesminister für Forschung und Technologie, 5th edition, Bonn, 1982, pp. 93–5.
21. Hagedoorn and Prakke, *An Expanded Inventory of Public Measures*, pp. 30–1.
22. Central Office of Information, *The Promotion of the Sciences in Britain*, Pamphlet 140, London, 1978.
23. Rothwell and Zegveld, *Industrial Innovation and Public Policy*.
24. Zermeno-Gonzales, R., 'The Development and diffusion of industrial robots', Ph.D. thesis, University of Aston in Birmingham, 1980.

25. MacTaggart, K. D., 'Technical innovation in industrial production', Ph.D. thesis, University of Aston in Birmingham, 1981.
26. Scarbrough, H., 'The control of technological change in the motor industry', Ph.D. thesis, University of Aston in Birmingham, 1982; and Fleck, J. and Scarbrough, H., 'Labour–management relations and the introduction of industrial robots in the car industry', International Conference on Robots in the Automotive Industry, Birmingham, April 1982.
27. See e.g. Beynon, H., *Working for Ford*, Harmondsworth, Penguin, 1973; Wobbe-Ohlenburg, W., 'Der Einfluss neuer Produktionstechnologien auf die Struktur der Automobilarbeit', doctoral thesis, Göttingen, 1982.
28. See e.g. Goldthorpe, J. H. *et al.*, *The Affluent Worker*, Cambridge, Cambridge University Press, 1968.
29. Pöhler, W. (ed.), *Damit die Arbeit menschlicher wird*, Bonn, Verlag Neue Gesellschaft, 1979.
30. Bundesminister für Forschung und Technologie, *Ein Programm und seine Wirkungen*, Frankfurt, Campus Verlag, 1982.
31. Lederman, L. L., 'Technological innovation and Federal Government Policy', in Stroetman, K. A. (ed.), *Innovation, Economic Change and Technology Policies*, p. 187.
32. Domestic Policy Review of Innovation, US Department of Commerce, (a) Sub-Committee on Industry Structure and Competition; (b) Labor Advisory Committee; (c) Sub-Committee on Public Interest; (d) Sub-Committee on Federal Procurement Policy, Washington DC, 1978.
33. Ibid. (a), p. 4.
34. Ibid. (a), p. 6.
35. Ibid. (a), p. 7.
36. Ibid. (a), p. 21.
37. Steward, H. F., 'Public policy and innovation in the drug industry', in Black, D. and Thomas, G. P. (eds), *Providing for the Health Services*, London, Croom Helm, 1978, p. 132.
38. Steward, F. and Wibberley, G., 'Drug innovation—what is slowing it down?', *Nature,* 284 (1980), p. 118.
39. White, B., 'State intervention in technology in the post-war years: an assessment of strategies, techniques and outcomes', Ph.D. thesis, University of Aston in Birmingham, forthcoming.
40. Drath, P., Gibbons, M. and Johnston, R., 'The supercomputer project: a case study of science, government and industry in the UK', *Research Policy,* 6 (1977), pp. 2–34.
41. Braun, E. and Macdonald, S., *Revolution in Miniature*, Cambridge, Cambridge University Press, 2nd edition, 1982, p. 32.
42. Drath, P., 'The relationship between science and technology: university research and the computer industry 1945–1962', Ph.D. thesis, University of Manchester, 1973; see also Levington, S. H., *Early British Computers*, Manchester University Press, 1980.
43. Braun, E., 'The role of universities in industrial innovation', Six Countries Project on Innovation, Stockholm meeting, May 1982.
44. Schwarzkopf, A., 'National Science Foundation experiments in industrial

innovation', in Maier, H. and Robinson, J. (eds), *Innovation Policy and Company Strategy*, Laxenburg, International Institute of Applied Systems Analysis, 1982, pp. 133–44; Duncalfe, M., *The Planning Implications of Science Parks*, dissertation, Polytechnic of the South Bank, London, 1982; O'Leary, M., 'Profile of the Innovation Centre, Limerick', 1981.

45. Gummett, P., *Scientists in Whitehall*, Manchester University Press, 1980.
46. Pavitt, K., *Technical Innovation and British Economic Performance*, London, Macmillan, 1980, p. 45.
47. Ibid., p. 100.
48. *Review of the Framework for Government Research and Development*, Cmnd 7499, London, HMSO, 1979.
49. Melman, S., *Pentagon Capitalism*, New York, McGraw-Hill, 1970; Pavitt, *Technical Innovation and British Economic Performance*; and Dickson, 'The Influence of Ministry of Defence Funding'.
50. Dickson, K., 'The influence of Ministry of Defence funding on semiconductor research and development in the UK', *Research Policy*, 12 (1983).
51. Braun and Macdonald, *Revolution in Miniature*, pp. 142 and 143.
52. Müller, J. et al., *Direkte und indirekte sektorale Auswirkungen der Investitionsaufwendungen der Deutschen Bundespost für Fernmeldeanlagen auf die Produktion und die Beschäftigung in der Bundesrepublik Deutschland*, Berlin, Deutsches Institut für Wirtschaftsforschung, 1982.
53. *Policies for the Stimulation of Industrial Innovation*, Analytical Report, vol. I, Paris, OECD, 1978, pp. 80 ff.
54. Allen, T. J. et al., 'Government influence on the process of innovation in Europe and Japan', *Research Policy*, 7 (1978), pp. 124–49.
55. Rubenstein, A. H. et al., 'Management perceptions of government incentives to technological innovation in England, France, West Germany and Japan', *Research Policy*, 6 (1977), pp. 324–57.
56. Jones, P. M. S. and Willett, J., 'Evaluation of the benefits of laboratory research and information services', *Research Policy*, 6 (1977), pp. 152–63.
57. *Pollution Control: Progress and Problems*, Cmnd 5780, Royal Commission on Environmental Pollution, Fourth Report, London, HMSO, 1974.

6 Market Control of Technology

In the previous chapter we discussed government policies for technology but, important as these are, the truly dominant influences
which shape technology in a capitalist or mixed economy are market
forces. Government policies determine the rules, shape the ambience
and occasionally take a stake, but the name of the game is survival
in the competitive production and sale of goods and services.
Financial and industrial managers will invariably claim that the
name of the game is profit. This seems as necessary to them as the
Credo is to the faithful, but no doubt Galbraith is right in putting
survival as the first requirement for the operation of firms.[1] Undoubtedly also a commercial enterprise is not a charity and must,
in general, reach at least some level of profits to continue in business
and in that sense the genuflexion toward Mammon is justified.
The role of market forces in the control of technology is to be seen
in competitive pressures which force firms to adopt certain technologies in order to survive and, additionally, the creation of opportunities for the growth of new firms or old.

In the following it will be useful to distinguish between the products of technology which serve the satisfaction of end user needs,
essentially consumer goods and services, and technologies which
serve the production of goods, essentially investment goods. The
forces which shape these two categories are sufficiently different
to warrant a separate discussion.

THE MARKET IN CONSUMER GOODS

The producer of consumer goods depends on the willingness and
ability of the general public to purchase his wares. In theory the
consumer has perfect knowledge of all the goods available on the
market, including their performance characteristics and price, and
chooses to purchase those goods which offer the greatest utility.
For each individual the utility is determined not only by the properties and price of the goods on offer but by individual needs,
means, priorities and preferences. The utility is, of course, determined *inter alia* by previous purchases and we speak of marginal

utility as 'the extra utility added by one extra last unit of a good' and of a law of diminishing utility, as 'the marginal utility of the good . . . tends to decrease.'[2]

In reality, the market is imperfect in many ways: the consumer is not in possession of either full or necessarily reliable information, the manufacture of some goods is in the hands of near monopolies or oligopolies, various political forces distort the pure market, including state intervention and the concerted activities of organised labour and/or employers. Despite all these market distortions— and there is no normatively negative implication in the term distortion—market forces are extremely powerful and the approximation of a perfect market does adequately describe many features of the real market. The essence of the market for our purposes is the unfettered choice of the consumer which determines the survival of a product and, ultimately, the firm. In some ways the consumer, within the limits of his or her income, is truly sovereign and may choose to buy or not to buy a given product at a given price. The manufacturer must adjust his strategy accordingly and offer a range of products which will ensure his survival.

But how do consumers make their choice? The first constraint is the total budget available. The second constraint is formed by necessities—things that must be bought to ensure the physical survival of the individual. Within this category there is a degree of choice, but the obvious one of 'take it or leave it' is eliminated. Wholemeal bread can be substituted for white bread or for potatoes, electricity for gas, an eiderdown for a blanket, but one or the other of these essentials must be bought.

For less essential items the consumer's choice includes buying or not buying and is not restricted to substitution. There is a whole hierarchy of needs—from what is truly essential for mere survival, through what is regarded as the bare minimum necessary in the given culture, to the niceties of life and right up to unashamed luxury. Both poverty and affluence are culturally determined— only the extremes are well defined and absolutely recognisable.

Underlying all the selection criteria for the purchase of individual products lies a spectrum of needs, desires and wants which are ultimately driven by some fundamental human attributes. Although technology and affluence are the result of social cooperation, they also serve the opposite need, independence and self-sufficiency. Similarly technology, the product of organised forms of social existence, serves the individual to boost his hierarchical position and to give him a competitive edge either by the mere fact of greater possessions or by the practical utility of such possessions.

We may distinguish several classes of needs which the individual, rather than society as a whole, seeks to satisfy by technology. The first and foremost need is that of survival, the satisfaction of the basic requirements of food, shelter, clothing, health and personal safety. As society becomes more complex, some of these needs become socially organised and become, in part at least, a social technological requirement rather than an individual one. Thus personal safety is now largely catered for by society and other basic needs are supported by a large social and technical infrastructure. Even the individual purchase of a loaf of bread sets in motion a vast wheel of social organisation and technical systems. The technical and organisational means of satisfying even simple individual needs have become highly complex, but the individual still seeks to fulfil them by individual purchasing decisions without worrying too much about the large number of levers his simple decision to purchase may have set in motion.

The fact that technology has given the individual a large degree of choice on the means of satisfying basic needs has added a super-structure of additional satisfactions obtainable with the basic ones. Satisfying hunger can now be a veritable feast of the palate—as in the past it could only be for the very rich—accompanied by feasts for other senses as by using beautiful china or listening to an accompaniment of recorded music.

The second driving force for individual purchases is a desire for autonomy and self-sufficiency. Perhaps paradoxically, in a society ever more complex and interdependent, the individual can use technology to attain a degree of apparent independence by the purchase of some technologies.[3] The washing machine makes the household independent of the services of a washerwoman or a commercial laundry, the deep-freeze gives an impression of autarky and independence from continuous external supplies, the motor car gives individual mobility independent of the functioning of a network of public transport. Of course this kind of autarky is more apparent than real, for it depends on the functioning of all kinds of systems, but the degree of autonomy obtainable appears to be worth the price paid.

Unhappily, the individual desire for autarky poses a social dilemma. For as more and more people make themselves independent of, say, public transport, so the public transport system decays and the motor car infrastructure grows, making the social choice of public transport increasingly difficult or impossible. The individual optimisation of expenditure adds up to a social disutility; Adam Smith's famous invisible hand fails to look after the social interest. Because motor cars interfere in their use with each other and, in any case, many citizens are unable to have or use one, the apparent freedom obtained by the motor car turns into an expensive illusion without escape.[4]

The third category of requirement which technology fulfils might be called the need for play or the desire for the good life (at least those aspects which technology can supply) and less hard work. There are numerous examples of technologies which provide amusement, support creative activities, provide direct pleasure, satisfy curiosity or combine these atributes with the fulfilment of vital practical needs. Examples in this category abound: television, cameras, electronic games, sports equipment, personal computers and information systems, lawn mowers and thousands of other items filling a technological Aladdin's Cave.

We may thus distinguish three main motivations for the purchase of products of technology: fulfilment of basic needs, attainment of autarky and sheer fun. There are several subsidiary driving forces and, in any case, most individual purchasing decisions are based on mixed motives. Among the subsidiary forces which drive the desire for technology we may single out two: first, the fact that technologies tend to form systems and the purchase of one item often leads to the need for the next. Examples include items required for the maintenance of the original equipment, as well as extensions, improvements and consumption items, whether they be fuel for the lawn mower or records for the record player. Because of continuous technical developments, obsolescence takes a heavy toll and old technology tends to be exchanged for new. Many purchases are not prompted so much by the desire to possess a certain product, but by the wish to have the latest, most modern, which may be more efficient or of improved specification.

The desire for the latest is related to the second subsidiary force leading to the purchase of goods, which is given by the ability of technology to serve, directly or indirectly, the enhancement of personal standing in a social hierarchy. Examples include the personal computer, which may give pleasure and amusement, may enhance the owner's prowess in computing and thereby give a competitive edge on the labour market, may improve the personal book-keeping and thereby offer tax savings, or may serve to enhance prestige by the mere fact of possession. There is an undeniable element of keeping up with the Joneses and one-upmanship in the possession of newly fashionable items. Ownership of 'top of the range' items does more than give the owner the best technology can offer; to have a big luxury car serves an image and prestige purpose as well as giving comfortable reliable transport. A sports car is not so much a conveyance of people as the conveyor of a message of dashing youthful zest.

These considerations of purchasing motivation bring us directly into the realm of marketing, advertising and symbolic imagery, where technology and its products move a long way from their simple stated purpose of fulfilling people's material needs. In theory, advertising serves the purpose of informing the public of goods and services available on the market, so that they can make their informed choices. The huckster has been replaced by the sales manager, and the sales organisation has been adapted to serve markets which have long outgrown the village square. In an abstract and world-wide market place the advertising agent has inherited the role of the market crier. Mythology has it that the consumer, fully informed by advertising and other marketing methods, governs technology by his purchasing decisions. Reality is somewhat different.

Highly selected and incomplete information is beamed at the consumer in an attempt by each advertiser to capture a larger market share and/or to create new needs. Very little is said in advertising about the satisfaction of basic needs—for these are usually satiable and are well below the level of economic activity which the advertiser attempts to stimulate. Autarky is stressed and so is the good life and the pleasures derivable from this or that purchase. Yesterday's technology is implicitly useless while today's will remain at the top for ever and a day. The lucky purchaser of a caravan can roam the earth unfettered by crowded sites, restrictive regulations, sanitation problems, or any other intrusions of reality. Bliss is guaranteed to those who are discerning enough to buy a washing powder which makes their washing not only whiter than that of their neighbours but also whiter than white. Happiness and delight are readily available off the shelves of the advertiser and come in every shape that modern technology sees fit to present itself in. Competition between many virtually identical goods is reduced to competition between advertising copy—it is not what you are selling but how you are selling it. The net result of constant stimulation of the appetite is a constant feeling of want—an insatiable craving for goods and a permanent sense of disappointment when the purchase of happiness does not bring ecstasy in its wake.

There are several forces at work to curb the powers and excesses of the advertisers, though the countervailing forces are but Davids against the mighty Goliath and their slings lack the magic power of their biblical predecessor. The curbs are legislation on the one hand and 'consumerism' on the other. The latter is an ugly word to describe a miscellany of movements designed to provide objective testing and information about goods and to protect the consumer against falling into the many pits dug for him by modern sales methods. The movements vary from local to national organisations and from the humble to the sensational. Ralph Nader is probably the best known practitioner in this field. Numerous magazines, such as *Which?*, fight quite successful rearguard battles against the ever attacking sales forces.

Governments, nominally the guardians of the public interest, intervene with varying degrees of determination and success. It is now mostly forbidden to make blatantly false claims in advertisements, goods have to be correctly labelled and described to a prescribed minimum standard of information. These controls are additional to those on weights and measures which are among the oldest technical functions of government and additional also to controls on the composition of goods for sale, e.g. on food additives or the safety of children's toys. Unfortunately governments feel bound by many interests and tend to license commercial television and, horror of horrors, even unbridled cable and satellite television, while wagging a feeble finger at the advertising merchants warning them to be on their best behaviour, or else suffer the consequences.

To sum up, technology is driven from the consumer demand side by several requirements which consumers express in their purchasing decisions: (i) the satisfaction of basic needs; (ii) a quest for autarky and hierarchical position; (iii) the desire for play, pleasure and comfort. Many products of technology fulfil several requirements simultaneously and particularly the satisfaction of basic needs is now usually coupled with the provision of pleasure and creature comforts. Even consumer technologies tend to form systems, so that the purchase of one item tends to favour the purchase of supplementary items. Built in obsolescence and social competitive pressures are further incentives to buy. Whereas fundamental needs can be satisfied and satiated, many products of modern technology create insatiable demands. Particularly positional goods can never, by their very nature, reach saturation, for it does not matter what you own but what position in the stack you occupy.

Humans are driven by many needs and desires and technology is very much better able to satisfy some of them rather than others. The tragedy of technology is that it can fulfil needs of autarky, but is unable to fulfil needs of cooperation and belonging. This is even more tragic as technology initially developed out of cooperation and is perhaps its crowning achievement. But the relentless drive for higher efficiency and higher production, achieved largely by division of labour and automation, have removed much personal cooperation from the sphere of technology and replaced it by impersonal procedures—*la technique.*[5] In private life technologically-enabled autarky has led to increased isolation and excessive stress on remaining relationships. The family sitting in silence before the television screen has become a stereotype of our time, as the television programme takes the place of active social intercourse. Even the friend needed to play a game has been rendered obsolete by the electronic game—the chess-playing computer instead of the mate.

When we speak of changing values in modern society, we do not really mean changing values but the realisation that technology cannot cater for the oldest values people hold—the sense of belonging, friendship and group allegiance. To channel these feelings into benign activities is the great task of community and political leaders, a task in which technology hampers rather than helps. The great task of technology policy is to foster those aspects of technology which help community life, while a prime objective of general policy must be to foster cooperation and replace lost social forms by new ones.

Although individual purchasing decisions do control the market and thereby the direction of technology, individual sovereignty is severely limited by several factors. First and foremost by the obvious limitations of cash—the individual can only purchase up to the limits of his disposable means. The second limit is that of information. Not only is it difficult for the consumer to fathom the intricacies of performance specifications, he is also flooded by enticing promises masquerading as information. In fact so much has been promised and so many hopes raised that some disillusionment with technology has become inevitable. The third limit on consumer control of technology is imposed by the fact that choice can only be exercised among what is on offer. Consumers determine—with all the constraints upon the exercise of their free will—which and what quantities of those goods offered for sale shall be sold. Suppliers determine what shall be offered. Long lead times and enormous investments necessary for many products of modern technology make it vital for producers that they be highly selective in their offers. The vast productivity of modern production and distribution methods militates against too wide a range of choice. The famous adage of Henry Ford 'they can have any colour they want as long as it is black' no longer applies to colour or other minor features, but it certainly applies to the basic characteristics of most mass produced goods. Add to this the oligopolistic tendencies in many branches of industry and the paradox of decreasing choice among an increasing range of goods becomes apparent. We may buy colour television and cars and cameras and a myriad of other miracles of modern technology, but we have to be pretty sophisticated to tell the difference between one make and the next except in their advertising copy.

Thus technology fosters the herd instinct without strengthening cooperation; people lose individuality without gaining in corporate spirit. In fact the herd instinct imposes yet another limitation upon the freedom of choice in matters technological. Not owning the latest gadget, once it has acquired some ubiquity, means a degree of exclusion from the community. Once enough people have acquired a colour television set, then the lack of one is either a sign of perverse isolation or of lamentable poverty. These tendencies are reinforced by the decline of older systems which previously fulfilled a similar role. This is particularly blatant in the case of the motor car, but is also apparent with information systems preceding television. Further reinforcement is given by what remains of social intercourse, as this tends not only to revolve round the latest technologies and their offerings but also assumes their ownership as a matter of course.

The general acceptance or failure of new consumer oriented technology is a fascinating subject which seems inadequately explored. Undoubtedly a new technology must offer sufficient advantages, at an acceptable price, over its predecessor to have any hope of acceptance. Stereo record players succeeded mono-players, whereas quadraphonic sound was rejected, despite all sales effort. Sufficient advantage over predecessors, fulfilment of one or more need/desire categories, few teething troubles—those are some of the necessary conditions for public acceptance of a new technology. But what sufficient conditions are remains one of the mysteries of modern civilisation and the prize of unravelling it seems beyond human grasp.

With modern mass production and mass marketing methods and instant feedback on consumer preference from electronic check-out points there are real dangers of declining consumer choice on the one hand and market instability on the other. The very concept of the market depends on a range of choice and on the possibility of some producers winning relatively minor advantages over others and the latter either going out of business or adapting to prevailing conditions. If the development of a new product takes years and millions of whatever currency of investment and the fate of a huge firm with tens of thousands of employees depends on its success, then the cosy concept of a market becomes untenable. For established products for the mass market, competition is shifted into the realm of verbal and pictorial warfare among advertising agents. True competition takes place among smaller producers of specialist or novel products and among a small range of truly innovative ideas which large firms launched by way of diversification and to spread their risks.

The law of diminishing returns for further development of already highly developed technologies operates ruthlessly. To get any further improvement out of an internal combustion engine, or a jet engine, or an established chemical process, absorbs vast effort and brings but small improvements. Yet somehow even oligopolistic firms persist in their efforts, often supported by governments, and the total economic benefit of such efforts must be in grave doubt. If one firms succeeds, the others might not survive and therefore they all do it. Market competition assumes that the departure of a few producers does not alter the total production system, while obviously the departure of any oligopolistic firm has such dramatic effects that modern governments simply cannot allow them to disappear. The corner shop may sink without trace, yet Chrysler and British Leyland may die a thousand commercial deaths and still float along on a wave of subsidies and guaranteed loans. The consequences of their closure simply cannot be contemplated.

Thus competition forces firms into frenzied and perhaps economically unjustifiable expenditure on development of slightly improved products, and governments feel obliged to hold a safety net under those firms who fail to hold their own on the commercial tight-rope. This measure of last resort does not diminish competitive pressures, for nobody likes to fall even knowing there is a safety net. Indeed if anything it spurs on competition for without some minimal re-assurance firms might not embark upon some of their more foolhardy exploits.

The relentless pressure to squeeze the last ounce of further improvement out of a mature technology is, no doubt, a result of competitive pressures. Yet it is hard to escape the conclusion that at least some of the pressure comes from within rather than from without. High technology firms, employing large numbers of highly motivated scientific and engineering personnel, would find it very hard to call a halt to development work once the latest model has left the pilot stage. Continuous development, ever new models, are part of the make-up of the firm and the expectations of its management and human capital. If the firm dispensed with its scientists, it would commit to death its known shape and structure —a kind of suicide only despair can induce. The scientific manpower of a high technology firm forms the major part of its existence and its self-image and the continuous quest for new solutions is ingrained in the very fabric of the firm. A high technology firm without R & D effort would be as Macbeth without his Lady or Faust without Mephistopheles.

THE PROVISION OF SERVICES

Consumer products can be packaged in convenient form for marketing. The consumer can contemplate the qualities, utility and price of the product and its rivals and, with all the limitations upon the operation of the market, make reasonable choices as to what to buy or not to buy. The goods come in readily discernible packages and the purchasing transaction is a simple transfer of ownership from seller to buyer. The transactions are made the simpler as the goods are truly man-made (apart from the content of non-renewable raw materials) and can be infinitely repeated, so that no moral issues of true ownership or the purchase of privilege arise. The purchase of land provides a counter-example, as land is not man-made, cannot be replenished and the possession of any plot provides a unique and inimitable combination of advantages.

Some products, particularly means of transport, have the unfortunate property that even the legitimate use by one owner interferes with that of another and thus the marginal utility for any purchaser depends not only on his or her decisions, but on those of others. Apart from that and from certain rules of safety and non-disturbance to others, the result of a purchasing transaction for a product of technology is the unfettered use of the product by its owner.

This is in sharp contrast to the purchase of a service. No right of ownership is conferred by the purchase and some services at least are not readily packaged for the market. The consumption of a service is a temporal affair. The holiday is soon a mere memory, the journey ends on arrival, the meal is finished with the last bite, and the show is over as the curtain falls. This kind of service is quite easily packaged for the market and can operate smoothly in a market economy. Instead of paying for the ownership of a product, the customer pays for services rendered.[6]

Nevertheless, several major problems arise in the provision of services in a market economy. The first difficulty is that of catering for minority interests. No modern opera house can survive without subsidy, as the cost of providing the service is higher than the amount the limited clientele is willing or able to pay. This problem is very similar to that of providing goods for minority interests or needs. The provision of drugs for rare diseases or technical aids for unusual disabilities is impossible without some financing mechanism which bypasses the operation of the market.

The problem is made infinitely more complex by competition from consumer products. As the consumer can become self-sufficient in his transport needs by buying a car, so the provision of public transport is rendered increasingly less economic. The availability of television and video recorders makes the consumer less likely to go to the theatre or cinema and causes added economic difficulties for these services. The competition from consumer goods which cater for autarkic needs and make the individual independent of services has prompted Gershuny to postulate a self-service society.[7] Do-it-yourself tools and devices replace the decorator and carpenter: frozen food and microwave ovens take customers away from restaurants. A vicious circle sets in as fewer customers mean a more expensive and worse service, thus driving more customers into an autarkic mood. In this way even services which might have catered for a mass market are driven into the position of services catering for minority needs.

Even greater problems arise in the provision of services which are socially vital but would be economically inaccessible to large classes of potential users. Educational services are vital to the functioning of a modern technological society, yet many families could not or would not contemplate paying the full cost of provision. The state intervenes for a variety of motives—to ensure an adequate supply of skilled manpower, to ensure uniformity of standards of equipment and achievement, to enable everybody to participate, to keep control over curricula, to spread the cost over the whole population.

Similar considerations apply to modern health care or, more commonly, modern methods of dealing with the sick. As medicine has become more scientific and medical treatment based on more and more sophisticated technology, so the cost has soared. Many governments are now frantically trying to extricate themselves from their basic obligation to provide for those stricken by disease and to return medicine to the market place. A combination of private insurance companies and private medicine are to take over where the state leaves off, with the latter catering only for the poorest of the poor. Thus the progress of science appears to drive society back to a non-caring commercialism which is closely allied with the ability of technology to make the individual independent of help and cooperation. Even where the sick individual needs help, this is to be sold in a quasi-market by monopolistic suppliers to consumers who are driven by sometimes desperate need and have neither information nor choice. Because the state finds it difficult to finance a technology-driven medicine, this is forced or allowed to tread the market path so apt for technology and so ill-suited to health care. Part and parcel of this trend are modern attempts to market self-diagnostic devices, thus helping people further on the lonely path to autarky.

A similar fate befalls many social services. As more people are either economically excluded from the benefits of technological society or find an existence sustained by machine and gadget too lonely to be acceptable, so they swell the ranks of the very old or physically incapable needing caring help. The nuclear family and consumer society have made this help unavailable from private resources, and the modern state needs to step into the breach. Although technology can help with various aids, nevertheless the essence of care is human contact and thus social services are, generically, labour intensive. Severe problems of financing arise and there appears to be an acute need for finding new ways of raising the funds necessary to keep a caring society functioning. As productivity rises and markets for industrial goods become satiated, it ought to be possible to channel surplus resources into social activities so badly needed and so ill-provided for.

Many services are, of course, not marketed to the consumer. These are services to industry or government, including goods transport, specialist design, legal services, defence and, last but by no means least, administration. It is the latter which occupies such a vital part of the total economy as it deals purely with information. This is a curious commodity gaining notoriety as it is manipulated by the latest generation of fashionable technology—information technology.

Among the many slogans which are supposed to render social structures instantly transparent, currently the most popular is that suggesting that contemporary society is an information society.[8] If we are to believe that this oversimplification contains even a kernel of truth, then information must be at the very heart of our economic activity. Yet our understanding of basic economic relationships with regard to information is infinitely poorer than that for other subjects of economic activity, such as the production and marketing of manufactured goods or of primary products.[9]

If we regard information as a trading commodity, as in many ways we must, it becomes immediately obvious that it has some curious properties, of which we shall list but a few.

(i) information is difficult to define and covers a multitude of sins, from tables of logarithms to Shakespearian drama;

(ii) it may be hard to measure the cost of production of information, yet to measure the benefit or cost of consumption is well-nigh impossible;

(iii) as the value of output is impossible to define, so the efficiency and productivity of the whole information sector must remain undefined. In view of the fact that this sector comprises more than half the economy, lack of knowledge about its productivity might be regarded as a serious deficiency.

Although many classifications of information are possible, the most important classification divides it into just two classes according to end use:

(i)   information as an item of final consumption;
(ii) information as an input to other economic activity.

Examples of the first type would include artistic and entertainment activities. We read a novel for the pleasure of it; the 'consumption' is an end in itself. The second type clearly includes all information activities within a firm, such as planning, accountancy, personnel records, research, legal activities.

No classification can cater for the occasional chance utility of information: serendipity. It is this 'browsing' quality of our intake of information and the amazingly tortuous ways in which circumstances auspicious to a particular activity arise that make it impossible for any information system to be perfect. For we can, at best, only supply known needs; new needs arise out of new constellations. There is a kaleidoscopic, foraging quality about our intake of information, which runs counter to all attempts at designing perfect information systems.

We must, however reluctantly, reach the conclusion that a high proportion of information cannot be given an economic value—it is truly priceless—and yet it is such freely circulating information coupled with the imaginative observer which makes the cultural, social and economic wheels turn round. We must regard the free availability of a vast variety of information as a public good of utmost importance. If the world be a stage, then a kaleidoscopic shifting pattern of information is the back-drop needed to unfold the play.

The universal multi-faceted nature of information and its essential presence in all human activity make it impossible to obtain a general economic theory of information. Only some special kinds of information are the subject of trade and thus a proper object of economic consideration. In particular, information intended as an item of final consumption has a market value and is subject to market

forces. The value of such information is related to only one quality
—its popularity. If we examine the total market value of a book,
a song, a symphony, a poem, or a play, it is related purely and
simply to the price individual consumers are prepared to pay and
the number of such consumers.

Popularity is one way of expressing perceived relevance, a consensus on high relevance, and it is relevance which determines the
value of all information. Clearly the marginal utility or relevance
of knowledge of a source of potable water is infinite to the person
dying of thirst and very small to the average city dweller. The
marginal utility of information leading to a commodity is closely
related to the marginal utility of the commodity itself. We can
generalise from this commonplace example to say that the value
of information used as an input to an economic activity is closely
related to the value of the activity itself and to the importance the
information plays in it. If somebody expects to make a million
pounds out of manufacturing and selling a superior mousetrap,
a patent covering its design may be worth quite a lot to him, as may
be the knowledge of some quirk in the behaviour of mice which
makes the trap work. On the other hand, all information on the
mating habits of mice may be quite irrelevant and worthless, as may
be many hundreds of patents covering a variety of designs of mouse-
traps. This is blindingly obvious until we consider a major difficulty
—how is one to know which patent and which quirk of behaviour
are important? If we exclude the possibility of divine guidance,
two ingredients are essential—wide reading about mice and about
methods of controlling them and connecting two previously unconnected items of information, the patent and the quirk of behaviour
of mice.

Trivial as the above example may be, it highlights the difficulties
of making information relevant and therefore valuable. Not only
do we not know the utility of an item of information till after we
have consumed it; we often only know it when, by some chance
or creative act, we have brought the item of information into a
constellation of other factors in which it may play a key role. The
marginal utility of an item of information depends on circumstances
of time, place, conditions and other items of previously unrelated
information. We may have to absorb masses of irrelevant information in the hope of finding something we need, often not knowing
in advance what it is we need.

The price we should be willing to pay for an item of information
is related to our expectation of the economic return we may derive
from it. Unfortunately the utility is often not known in advance;

although we may know the need to absorb some kinds of information, say an instruction manual, the relevance of other kinds of information remains uncertain till after it has been consumed. The cost of consumption of information may thus range from a necessarily incurred cost to a highly speculative investment.

The reduction of uncertainty of relevance is the main task of information services and consultants. The most common situation for a consumer of economically useful information is the knowledge that what he needs exists somewhere, hidden among hundreds of thousands of words and a myriad tables. Sifting and preparing the information for a particular perceived need—possibly making some judgement on its reliability as well—is the economic activity of the information scientist. This task is made much more difficult by the sheer mass of information available. Computers were supposed to make the shifting so much easier, but have helped substantially only with some types of information. The cost of consumption of information is high and remains high despite the use of information processing technologies. Information suffers a high degree of congestion costs.

The mode of dissemination of information varies with the number of people to whom it is relevant and with the temporal qualities of the information. Thus general news is broadcast not only because it is relevant to a large number of people, but also because it is ephemeral—words writ on the television screen. Similarly relevant but more or less permanent information, the durable goods of the information business, is recorded in what we might loosely describe as handbooks: tables of physical constants, collections of laws and regulations, telephone directories, Who's Who and so on.

The aim of systems like 'Prestel' is to bring information of varying durability, relevant to large numbers of people, onto the television screen. If updating needs to be done fairly frequently, it is much cheaper than reprinting handbooks. If the information is truly readily available through clever algorithms, then its use is more convenient than that of handbooks normally found only in libraries, and a larger range of information of varying degrees of fleetingness, from news to theatre programmes to railway timetables to dictionaries, can be incorporated. The economics of a selective broadcasting system depends on a critical balance between the cost, the convenience of use, and the perceived relevance of the available material.

Computer information systems need not, of course, work in conjunction with a broadcasting system. Indeed perhaps the most important applications are within commercial organisations where stock control, accounts, lists of parts, machine scheduling, patent files, design data, and much other information are held on computer and made accessible on display devices within the organisation, possibly in widely dispersed locations. The economic value of this type of activity is extremely hard to gauge. Because the information is not an item of final consumption and because the quantity and quality of information required are indeterminate, the economic value remains indeterminate. A balance must be struck between such items as: how much time is saved by providing the information in one form rather than in another; how much value does the information add to the final product; what is the cost of collecting, holding and disseminating the information; what is the cost of unnecessary use of the information; what is the degree of certainty of relevance; is serendipity catered for and, if so, at what ratio of actual cost to likely benefit? The possibilities of answering these questions range from difficult to impossible; yet we must know these answers if we are to otpimise, in the economic sense, an internal information system.

Even the systems of the kind described above will cater mostly for information needs in simple factual categories. The main problems concerning this type of information are ease of access through suitable classifications and algorithms, cost effectiveness and catering for serendipity. Certainty of relevance can be high for information with high factual content. The more verbal and speculative the information becomes, the more difficult it is to rank into relevance. This point is crucial, for the marginal utility of incurring the costs of consumption of information is proportional to its relevance. Hence so much effort is devoted to computerized access to types of information which are not readily amenable to this treatment. Can we really guess the relevance of a doctoral thesis from a pathetic little list of key words?

THE MARKET IN CAPITAL GOODS

Although it must be the ultimate aim of technology to produce consumer products, much of its effort is devoted to producing goods which serve the production of consumer products and the maintenance of the infrastructure of technological society. In a sense technology devours its own children—in order to produce what the consumer desires it consumes much of what it produces.

While market control of consumer products is exercised by the purchasing decisions of millions of lay consumers, the market for investment goods consists of a much smaller number of expert buyers. Capital goods range from enormous machines such as turbo-generators or aircraft—with single investment decisions over many millions of dollars or pounds sterling—to small tools not dissimilar to those the housholder buys. There is another category of goods, neither capital nor consumer, yet one that plays an important role in total economic activity. These are semi-finished or intermediate goods (or services) which manufacturers buy in for incorporation into their own final products. An aircraft consists of thousands and thousands of parts and sub-systems, of which a very large pro-portion will be supplied to the aircraft manufacturer by sub-contractors. A large fish is surrounded by many small fish all living in a symbiotic relationship. The industrial system is nothing if it is not a system—one of many interlinking units supplying and servicing each other.

Disregarding intermediate products and disregarding even capital goods which serve the provision of a service, such as aircraft, for our present purpose, we shall concentrate on the market control of capital goods which serve the production of other goods. We thus concentrate on production technology and its use.

Industrial management, charged with the task of ensuring the survival of the firm, is driven by several motivating forces. Its basic purposes can be summed up easily: to ensure the efficient pro-duction and sale of its products in a competitive environment. Thus efficiency of production—or high productivity—is a fundamental goal of management. The secondary goal, deemed essential to the achievement of the first and to the fulfilment of the role of manage-ment, is control over the total activities of the firm, including, of course, the production processes. It has been argued, notably by Braverman,[10] that control is the prime aim of industrial capitalism subservient to the profit motive only. The argument is somewhat unconvincing as control must be an ancillary aim of any body of persons charged with the achievement of any given output or activity of any organisation.[11] The underlying purpose of any industrial enterprise, irrespective of ownership, is the efficient manufacture of its products. Any society which organises production in order to attain the goods produced, rather than mainly in order to provide work for its population, must strive after efficient production and any management charged with achieving this aim must seek to con-trol the process of production. Only if the work process achieves the status of an aim in its own right rather than that of a means of achieving efficient production, could the forces driving manage-ment change. At present at least, the *raison d'être* for a manufactur-ing unit is the production of goods. Until other aims are explicitly declared, management will strive to control the factory in such a way as to achieve maximum possible efficiency of production within the constraints it finds itself in.

The productivity management attempts to achieve is the so-called factor productivity, which is the ratio between total output and total inputs. At any given moment the inputs consist of buildings, energy, machinery, bought in parts, materials and labour. It is customary to subsume everything other than labour under the term capital and simply speak of two factors of production—capital and labour. As the costs of these two factors change, a different mix is employed. In recent years the factor labour has become significantly more expensive and strenuous efforts have been made to increase labour productivity. There are many ways in which this can be achieved—by increasing the capital to labour ratio, amplifying the efforts of each worker by more machinery; by increased division of labour on the classical model of Adam Smith; by increased supervision and quality control or by other organisational means. Because of the great cost-effectiveness of microelectronics it may now be possible to increase automation even without increasing the ratio of capital to labour.

When all is said and done, the decision by management to buy an item of production equipment will be determined by their perception of whether or not the equipment will serve their purposes. The kind of questions that will be asked implicitly or explicitly, include the following: will management control over the work and production process improve; will labour productivity increase; will the equipment be cost-effective in increasing capital productivity; will it quickly repay its investment in terms of increased overall productivity; will the quality of the product improve, will the image of the firm be enhanced?[12] The list is not exhaustive, yet it includes some questions which cannot be rationally answered, such as the firm's image. But even for those questions that, in principle, have a rational answer, this is not always obtainable and the management decision is an internal political process, underpinned by rational argument, but based also on hunches, judgements and internal power relations. Questions of acceptability of a technology to labour and problems of training, recruitment and maintenance may also play a crucial role in investment decisions.

If the aims of management can be traced back to two fundamental motivations, so can the aims of labour. They share the aim of ensuring the survival of the firm but, unless consultation and co-determination procedures are highly effective, do not feel that the responsibility for ensuring survival is theirs. Thus this aim tends not to be an explicit part of the labour repertoire and instead becomes a component of the 'them and us' relationship between management and workers, although the explicitly stated aim of ensuring security of employment does, of course, imply an interest in the survival of the firm. The main explicit aim of labour is to obtain the greatest possible share of any benefits accruing to the firm through any changes in organisation, markets, products or production technique. Secondary aims include the securing of maximum control over the work process by the workers and the achievement of the best possible working conditions. The degree to which these fundamentally divergent aims of management and labour can be reconciled or brought into congruence determines the tone of labour relations in a firm and thereby probably the fate of the firm.

The purchasing decisions on production equipment, made by management, are the market decisions influencing the course of technology in this section of the capital goods sector. Such purchases are made for a variety of motives and in response to different needs, such as: (i) normal writing off and replacement of worn-out machinery; (ii) premature replacement of machinery because of successful innovation in the relevant production technology; (iii) acquisition of additional machinery as previous mechanisation or automation 'gaps' become closed by innovation or shifts in the relative cost of capital and labour; (iv) enlargement of capacity; (v) changes necessitated by process innovation; (vi) changes caused by product innovation; (vii) changes in costs of materials or energy; (viii) other external or internal changes of circumstances.

In essence, management will attempt to remedy weaknesses in the production chain and these may be caused by all the factors listed above. Some weaknesses are indeed caused by faulty machinery as for example the imprecision of a worn out machine tool. Other weaknesses may be caused by the new availability of a superior technology, which makes a perfectly sound machine or process obsolete. Some weaknesses are purely organisational and can be remedied by such means, others are purely technical, many consist of a mixture of the two and the remedy consists of new technology coupled with new organisational forms.[13]

For a variety of immediate reasons and with mixed immediate motives, management will attempt to direct its purchases of capital goods in such a manner as to comply with its basic aims of efficiency and control. Innovation in the capital goods sector occurs either in response to new needs caused by process or product innovations (demand pull), or by creating a need by catering for basic aims of management better than the predecessor technology (technology push). Currently microelectronics is creating a demand for itself by closing many automation gaps, thereby offering great improvements in labour productivity, and by offering better control both in the physical and managerial sense.

The activity of introducing new manufacturing procedures, without changing the fundamental process of manufacture or the product, has been termed manufacturing innovation. Currently much of this innovation involves the introduction of electronic and computer controls over processes which were previously controlled by different means. In particular the introduction of electronics to replace previously human control—automation by electronics—causes much heart-searching.

The prime example is the introduction of computer numerically controlled (CNC) machines—machine tools whose operation is controlled by a programmed computer instead of by the direct manipulations of a skilled operator. The machine can do everything a lathe or other machine tool can do, with quite a bit added. The operation is fully controlled by a computer programme fed into the machine and it will turn out the desired parts quickly, accurately and reliably. Though the operation is not as fast as that of a fully automated machine purpose-built for the mass production of some part, the CNC machine is much more readily programmed and admirably suited to the production of small batches, which were previously made by hand-operated machines.

There are at least two ways in which the work process can be organised round a CNC machine. The machine can be programmed by a white-collar worker in the design office, possibly in conjunction with computer aided design (CAD). Maintenance and repairs are carried out by highly trained specialist maintenance workers who may also look after the setting and maintenance of the tools used by the machine. The remaining tasks of supervising the machine's correct functioning, feeding it with material and lubricant and removing the finished products can then be carried out by a pretty mindless machine-minder. The division of labour is virtually complete and all the important functions are carried out away from the shop-floor, thus giving management maximum control. All the skills required are concentrated on the maintenance and setting tasks on the one hand and the design and programming task on the other. Manual dexterity, eye-hand coordination, experience and feel for the job are replaced by abstract skills and theoretical or recipe prescriptions and these are concentrated in or near the management domain.

The alternative way of operating CNC machines is by training a skilled machinist to programme and maintain such a machine, or at least to fulfil a substantial part of these functions. In this case the programming, setting, machine minding and so on are all concentrated in the same hands, those of the skilled operative. He will not only bring his skill and experience to the new tasks, but will also take pride in the machine and operate it to best effect. Communications and cooperation between office and shop-floor may be improved and certainly more control will reside on the shop-floor. The division of labour will be less extreme and yet the two solutions sketched here seem to be economically neutral. At least the efficiency of the two methods appears sufficiently close as to show no systematically measurable difference in factor productivity.

The decision between these two modes of operation, or many possible variants, rests with management and will be made on the basis of attitudes and values, local political circumstances and the availability of skilled labour. The technology optimally required for the two solutions is subtly different and indeed it is possible to construct machinery more suitable for one solution rather than the other. This is a clear case where management philosophy and factory politics affect the course of technological development. The room for manoeuvre may not be large, but it is certainly significant and shows that technology is not autonomous, even within the constraints of the present system of production. It is possible to use technology in more or less humane ways and technology is capable of adapting to such requirements.[14]

It is quite wrong to assume that requirements of efficiency uniquely determine the direction which market forces compel technology to take. Efficiency of production, or factor productivity, is undoubtedly a major constraint on decisions about production technology and organisation. But being a constraint is very different from being a determinant; the constraint leaves a sufficient number of variables to be decided and sufficient room for value judgements and internal conflict. The outcome is a compromise—political in the sense of being the resolution of conflicts of interests and opinions. Efforts such as the German programme on 'humanisation of work' are much more than a sweetener of the bitter pill of modern factory work. They are, at best, a serious attempt to improve working conditions within the constraints set by the requirements of competitive efficiency of production and have had significant success.

As with the marketing of consumer products, the sale of investment goods requires a considerable infrastructure. Not only is there a need to acquaint potential customers with the capabilities of the goods on offer, but a range of support services is also necessary. Both NC machine tools and robots provide examples of the necessity for the manufacturer to work in close cooperation with potential users to develop the applications technology in detail. A robot manufacturer thus needs to know a great deal not just about the varied aspects of robot engineering, but needs to be fully conversant with spot welding or paint spraying or whatever application the robot is to be used for. It is only by communication of application needs and robot capabilities that a practical system can be developed. Without detailed applications knowledge the robot would remain a laboratory curiosity.

Because a revolutionary new production machine requires new skills to operate it, the manufacturer of the machine is often called upon to provide training courses for future operatives and maintenance technicians. It is the training role which opens up possibilities of feedback to machine design. Enlightened cooperation between users, workers and manufacturers can lead to machines which are not just more convenient to operate and more efficient, but actually allow the operative to use skills and experience and thus avoid the pitfalls of a polarised workforce where all knowledge resides in the office and the shop floor knows nothing but bitterness and frustration.

Major maintenance and repairs often have to be carried out by the supplier of the machines and this can cause awkward chicken and egg problems. For nobody wants to buy from a supplier with an inadequate maintenance organisation, while on the other hand nobody can provide such an organisation without sales to pay for it.[15] The building up of a sales, application engineering, training and maintenance organisation is a condition for the successful market penetration of a major technology and the very high cost of building up such infrastructure goes a long way toward explaining the slow initial start to sales of revolutionary machines. The same factors also make it unlikely for a large number of suppliers to succeed in any one market, as a sufficient flow of sales is needed to support an adequate infrastructure. Oligopolistic tendencies lie at the very root of the modern capital goods market.

The infrastructure and thereby the penetration of a technology can be aided by the formation of interest groups and by state support measures. If manufacturers and users get together to assemble and disseminate information and government supports research, training and initial introductions, the process of diffusion can be considerably accelerated. Robot technology in Britain provides a good example of this, with the British Robot Association and the

Department of Industry providing valuable services as joint midwives to the infant technology.[16]

The cosy image of technology being offered for sale on a free market and purchased by individuals attempting to optimise their own position, with all the purchases adding up to best common benefit, is clearly a romantic fantasy. We live in a complex world where sophisticated technology is produced by large organisations and the public interest too needs to be organised lest play gets too rough and injuries become too numerous and serious.

NOTES

1. Galbraith, J. K., *The New Industrial State*, Harmondsworth, Penguin, 2nd edn, 1974.
2. Samuelson, P. A., *Economics*, Tokyo, McGraw-Hill Kogakusha, 10th edn, 1976, p. 433.
3. Gershuny, J. I., *After Industrial Society?*, London, Macmillan, 1978.
4. Hirsch, F., *Social Limits to Growth*, London, Routledge & Kegan Paul, 1977.
5. Ellul, J., *The Technological Society*, New York, Vintage Books, 1964.
6. For a detailed discussion and classification of services see e.g. Gershuny, J. I. and Miles, I. D., *Service Employment: Trends and Prospects*, London, Frances Pinter, 1983. I am much indebted to the authors for making available an early draft of their book.
7. Gershuny, *After Industrial Society?*.
8. One of the first of a flood of publications in this area was Porat, M. U., *America: The First Information Society*, Comments prepared for the American Association for the Advancement of Science, Boston, February 1976.
9. For early authoritative comments on the economics of information see Lamberton, D. M. (ed.), *Economics of Information and Knowledge*, Harmondsworth, Penguin Books, 1971; some of the following comments have first been published as Braun, E., 'Some remarks on the economics of information', in Vasko, T. (ed.), *Telecommunications: Some Policy Issues*, Laxenburg, International Institute for Applied Systems Analysis, 1982, pp. 12–19.
10. Braverman, H., *Labor and Monopoly Capital*, New York, Monthly Review Press, 1974.
11. See e.g. Child, J., *Organisation: A Guide to Problems and Practice*, London, Harper and Row, 1977.
12. Bessant, J., 'Influential factors in manufacturing innovation', *Research Policy*, 11 (1982).
13. Braun, E., 'Constellations for manufacturing innovation', *Omega*, 9 (1981), pp. 247–53.
14. Wilkinson, B., *The Shopfloor Politics of New Technology*, London,

Heinemann, 1983 and Wilkinson, B., 'Technical change and work organisation', Ph.D. thesis, University of Aston in Birmingham, 1981; Sorge, A. *et al., Microelectronics and Manpower in Manufacturing: Applications of Computer Numerical Control in Great Britain and West Germany*, Berlin, International Institute of Management, October 1981; Council for Science and Society, *New Technology: Society, Employment and Skill*, London, 1981; see also 'The information society', paper by H. Rosenbrock given at EEC FAST conference held at Selsdon Park near London in January 1982; and Cooley, M., *Architect or Bee?*, London, Hand and Brain Publications, 1980.

15. Senker, P. *et al., Maintenance Skills in the Engineering Industry: The Influence of Technological Change*, London, Engineering Industry Training Board Occasional Paper, 1981; and Swords-Isherwood, N. and Senker, P., *Microelectronics in the Engineering Industry—the Need for Skills*, London, Frances Pinter, 1980.

16. Fleck, J., *The Introduction of Industrial Robots*, London, Frances Pinter, forthcoming.

7 Technology and Social Goals

Technology creates wealth, specific kinds of wealth, the products of technology. Indeed the creation of wealth is the essence of its existence, the *raison d'être* of all technology. In any society above subsistence level, wealth is defined by social consensus on what is desirable. Value is given by a combination of practical utility, the satisfaction of non-utilitarian needs and the acquisition of positional goods. Some societies are known to regard pigs as the main representation of wealth, in agrarian societies land is the main source of wealth, gold is wealth by ancient custom (although here wealth is somewhat confused with a means of exchange). Virtually all contemporary societies regard industrial products as desirable and therefore as objects of value and a source of wealth. This is a remarkable circumstance, worthy of some consideration.

The range of industrial products is so vast and they serve such varied purposes that it is difficult to see any communality of value. We may distinguish, however, in a rough and ready way, four types of utility and value provided by technology.

First, technology provides items of vital consumption and extends their range from what is truly needed to sustain life right up to sheer luxury. Apart from having provided many new options of fulfilling essential needs, technology has made possible the satisfaction of such needs for the vast populations inhabiting the modern world. It is technology which provides the essential food, clothing and shelter to sustain life for the untold millions which have inherited the earth and it is technology which enables many of them to exercise choice in their consumption and to add pleasure and well-being to mere survival. From the hovel to the palace, from the palliasse to the water-bed, from bread to smoked salmon, from a blanket to a mink coat—all these are now products of modern technology. Of course, many of these items can also be provided by craft methods, but the range of such products is substantially smaller and when a vast population has to be supplied cheaply with essentials, modern technology becomes indispensible.

Secondly, technology provides means of production: from the shovel to the combine harvester, from the hammer to the industrial robot, from the test tube to a vast petrochemical installation. Clearly, the simplest tools can be made by craft methods, but the more complex ones require the full arsenal of science-based modern technology and without these means of production, the totality of products—both in variety and in quantity—would be quite inadequate to supply the needs, let alone the wants, of contemporary vast populations.

Thirdly, technology provides the infrastructure necessary for the functioning of modern urban technological society. Again the range of products is as vast as the range of services rendered, all requiring industrial products. It would be iniquitous to mention particular services, but the main categories of service are supplies of water and energy, waste disposal, transport and communication, health, education and security. Sadly, the latter includes not just the fire brigade and the police, but also the most voracious and insatiable of all consumers of the products of up-to-the-minute technology—the armed forces.

Apart from the infrastructural services so obviously based on technology, all other services depend, in less obvious ways, upon technical equipment. Sport, theatre, legal and administrative services —they all consume products of technology and indeed are becoming more and more dependent upon them.

One of the main features of modern society is its interdependence through a vastly complex network of supplier–user and control relationships. Flows of goods, information and people form a pattern too complex to be grasped but wholly dependent upon products and methods of technology.

Fourthly and finally, technology provides what one might call aids to easy living. These range from devices which take the toil out of work that needs to be performed, say vacuum cleaners and washing machines, to toys such as cameras, video recorders, home computers or sailing dinghies. We lump together technical devices which do no more than ease the performance of tasks which are quite feasible with much simpler methods with those which cannot be performed without modern technology at all.

Some technologies do not fit well into any one of the above four categories of wants. The most obvious example is the motor car, this most ubiquitous and symbol-laden product of modern technology. On the one hand an important part of the transportation system and thus of the infrastructure, on the other hand a toy and major agent of easy living. The classification of wants can only be rough and ready as indeed it is a feature of modern technology to provide more than one motivation for its acquisition: means of transport must provide comfort and can be playthings; jet aircraft transport mail and holidaymakers; caviar provides food, pleasure and status symbol all at once.

To the extent that the fulfilment of needs by technological means can be separated into categories, it is the first and the last of our classes that make technology so highly desirable and highly prized. People all over the world, across all cultural and political boundaries desire the fulfilment of their basic needs and the easy life which technology can so marvellously combine. Some of the infrastructural items are also desiderata in their own right—things like telephones and mains water are nice to have—but the irresistible attraction of technology lies in providing cheap and plentiful supplies of essential and less essential items of food, clothing and shelter—and beautiful gadgets for an easy life. If there are doubts as to the value of those things, the doubts are most commonly expressed by the well-fed. For those who toil on the verge of subsistence the fruits of technology appear almost as the fruits of paradise; only those who spend most of their time playing with technological toys sometimes wonder at the futility and cost of it all.

That the products of modern technology are desired by many and thus form wealth cannot be doubted. But does technology provide what people desire, or do they merely desire what technology can provide? Langdon Winner argues the latter and calls the phenomenon 'reverse adaptation', by which the industrial system manipulates human needs in such a way as to satisfy its own requirements: 'beyond a certain level of technological development, the rule of freely articulated, strongly asserted purpose is a luxury that can no longer be permitted'.[1] It is true, of course, that major corporations manipulate the market and employ a huge array of subtle and not so subtle methods to persuade the public to buy their wares. It is true also that these corporations must produce very large numbers of identical or near-identical products in order to compete in an oligopolistic market and to justify their huge investment in planning, research and development, design and production equipment. Thus a product, once launched, will be force-fed to the public and any reluctance to buy spells disaster for the corporation. In Galbraith's words:

in addition to deciding what the consumer will want and will pay, the firm must take every feasible step to see that what it decides to produce is wanted by the consumer at a remunerative price. And it must see that the labour, materials and equipment that it needs will be available at a cost consistent with the price it will receive. It must exercise control over what is sold. It must exercise control over what is supplied. It must replace the market with planning.[2]

But to say that large firms are inflexible and use dubious means to ensure sales is by no means tantamount to saying that they do not supply what people need. People do not spend money on what they do not perceive as a need, despite all shady manipulations of wants. A world of free choice is infinitely preferable to one in which Big Brother tells people what they need. The political problem is to guard against excessive manipulation of wants—either by decree or by powerful sales organisations.

For needs to become meaningful, there must exist a potential of satisfying them. Needs in the abstract are either rather general or dreamlike; they become concrete only in the face of concrete offers of realisation. The articulation of real practical needs is a process of choice between solutions on offer. Leonardo's dream of bird-like human flight became the choice between Boeing and Airbus; the long quest for a cure for cancer becomes the choice between chemotherapy and radiation; and the offerings of computers are the human version of omniscience.

Whatever the foundation of wants may be, their satisfaction constitutes wealth and thus the function of technology is the creation of wealth in the shape of artefacts which people desire and are willing to work and pay for. But where wealth is created and accumulated, problems of distribution arise. In a true subsistence economy no large-scale inequalities can occur (although a few very wealthy individuals, such as landowners, may exist); for anybody who falls much below the average perishes. Where there is a surplus above subsistence, inequalities can and do arise. The further creation of wealth is used by the rich and powerful to become richer and more powerful, and in this sense technology is not neutral. It is on the side of the powerful, not because of its own peculiar nature, but simply because it is a creator of wealth and wealth tends to be accumulated by the wealthy. The fault is not specific to technology, but lies with social structures which allow an unequal distribution of resources, quite irrespective of the nature of the resource.

This simple view is, of course, insufficient. The problem is compounded and made infinitely more complex by the overall desire of society to increase its wealth. Society therefore feels a necessity to encourage the ever greater and more efficient use of technology and provides incentives to its creators and controllers. In part at least, often in a large part, these incentives are financial and tend to reinforce inequalities. Society seems willing to pay a financial tribute to those who advance its overall wealth. This is the meaning of the creation of monopolies by patents and of the myriad of other incentives to industrialists, entrepreneurs and members of the ancillary professions serving industry. Indeed it may be argued with some justification that when the incentive structure becomes distorted by non-technological interests, such as land speculation, the creation of wealth suffers. Similarly it may be argued that if the incentives become too narrow and crude, the social and industrial infrastructure suffers, to the detriment of social well-being. Investment opportunities which merely re-distribute wealth rather than create it are detrimental to material well-being. On the other hand excessive stress on the creation of material wealth may destroy social equity and damage the cultural structure which alone assures the well-being of non-material Man. Material wealth is only one aspect of society and unless it is used to underpin the emotional and the visionary, society becomes sterile and destructive.

Even in economic terms, products of technology are not the only wealth and technology is not the only creator of wealth. Indeed from the point of view of economics, everything that people are prepared to pay money for is wealth, whether it be a haircut, a book, a painting, a plot of land, a house, a motor car, a litre of petrol or a loaf of bread. The difference between the various categories lies mainly in the speed of consumption and in the degree to which the resource can be created by Man. Land can at best be improved by technology, petrol contains a large element of natural resource and a considerable component of technology, a book is almost wholly man-made. The proportion of natural and man-made ingredients are one characteristic of products and their value is determined by the scarcity of the one and the extent of the other. A painting does not in essence derive its value from technology, although technology is used in its production. Bread and motor cars clearly are products of technology, though agricultural products or natural resources enter into their value in a vital way. Thus wealth falls into many categories, with technological inputs varying from the marginal to the dominant and with speeds of consumption varying from zero, for land, to slow for capital goods and consumer durables, to very fast for consumption goods and services.

Services form a special category, as it is the consumption and its effect that matter, rather than the retention, albeit temporary, of a product. What matters is the viewing and hearing in the theatre, the shorter and better shaped hair in a haircut, the pleasure of effortless eating and the feeling of satisfaction in a restaurant, the fun of a cruise. Although most goods also serve ultimate purposes other than their possession, the money is exchanged for the goods and not for their ultimate effect. Yet in many services technology plays a major role and in nearly all of them at least a marginal one.

Although industry is by no means the only creator of wealth in an economy, it certainly is a major creator of wealth and industrial products are involved, to a greater or lesser extent, in most forms of consumption and in the underpinning of social structures.

The market value of products is given by their utility to the consumer and the price is fixed by the cost of production (as the lower limit) and the willingness of the consumer to pay. There are some goods, and they form an important category, where the price is almost unrelated to the cost of production and is determined mainly by scarcity. Gold is the prime example—although perhaps faith and convention play a greater role than scarcity. Similarly, the price of many rare natural resources or goods in short supply can be considerably higher than the cost of production and distribution.

A concept closely related to scarcity is that of positional value. Paintings of old masters, houses with rare views, membership of exclusive clubs are scarce, sometimes to the point of uniqueness, and their possession is reserved to those with the highest position in society, unless they are owned by the public. Hirsch has argued that, because of the value of positional goods, absolute wealth in a wealthy society is almost irrelevant to the individual, while relative position in the distribution is the all-important parameter.[3] For positional goods are always priced in such a way as to be accessible to the top social layer only and their possession is prized above all others. Positional value, as an expression of the ultimate aim for status, for having something most people do not have, is one driving force behind the constant quest for more economic growth and therefore for more technology. Even though technology creates positional values to a limited extent only, it is nevertheless commandeered to create more wealth in an effort to catch up with the elusive prize of possession of positional goods. It is a race with a moving finishing line, but products of technology are acquired on the way and some pleasure is derived from their acquisition.

The spectrum of human needs ranges from the necessities of life to its niceties. Technology can satisfy some of thes needs but in its very achievement causes the neglect and atrophy of other aspects of human nature. Technology has enabled countless millions to remain alive and a smaller but equally countless number of millions to enjoy varying degrees of luxury and of play. Technology has also provided the necessary infrastructure for all these activities to unfold and for social structures to fulfil their basic functions.

On the other hand, technology has disrupted much of social infrastructure and some aspects of technology interfere in their use with each other to such an extent as to limit their utility severely.

The motor car is the prime example of the latter problem. Each person acquiring a car interferes with the freedom of use of every other car owner. However, it is only the widespread ownership of cars which makes an adequate infrastructure of roads, traffic lights, filling stations and repair facilities possible. The social goal of achieving high personal mobility without congestion and without severe deprivation of the non-driver has proved elusive and can certainly not be achieved by the unrestrained operation of market forces.

The car is also the prime symbolic example for the fact that technology can cater for the autarkic needs of the individual but not for the need for cooperation and common activity. The lone driver in his metal box, unable to communicate, frustrated and hemmed in by fellow car users is a powerful symbol for both the utility and the futility of technology. Technology enables people to be self-sufficient and independent: their own transport, their own home entertainment, their own mechanized and automated household, soon their own information system. Technology also amplifies the individual's physical and organisational capabilities so that they can achieve single-handed what previously required cooperation of many people. The shifting of heavy loads, the mowing of a large lawn, the production of complex products on computer controlled mechanised production lines. . . .

Yet technology can do nothing for social, spiritual, cooperative, visionary and artistic Man. This does not mean that technology is bad, it means that we cannot put the totality of our faith in technology and that many facets of society must be developed without reference to technology. Its very success has tipped the balance—its very strength has driven people too far along the path where technology can show its prowess. Measures to strengthen cooperation and social cohesion must be taken which are independent of technology. The grave peril for society lies in the tendency for such measures to be highly dangerous and objectionable; mysticism, drugs, national and other fanaticism. Driving out the devil with the acid of Beelzebub.

Paradoxically, autarky and self-determination are not necessarily complementary, indeed they can be in conflict with each other. The kind of autarky which technology provides to the private individual does help his apparent self-determination, but makes him increasingly dependent upon the functioning of complex systems, including the maintenance of his equipment. Infinitely worse is the juxtaposition of autarky and self-determination in the factory or the office. Because each individual can command so much machinery, the functioning of the total system is dependent upon the individual's total conformity with narrowly prescribed rules. With few exceptions, the individual must perform according to strict procedures, his scope for own decisions is limited to the point of non-existence, as the vast complex machinery of the factory depends critically upon the fully predictable performance of each cog and wheel. So while the power of each worker is enormously amplified by machines, his decisions are severely circumscribed by his very power.

The political consequences of this paradoxical position are considerable. Peaceful relations and conformity in the factory must be assured, as the use of any individual's power against the system has dire consequences. Thus cooperation is as vital as ever, but it is a cooperation without individual commitment or individual scope; a cooperation of galley slaves bound together by a vast and complex machine which must be kept going. Cooperation can be assured either by social means of shared responsibility and co-determination, or by means of strict discipline enforced by fear of reprisals, or by performance incentives. Alternatively and additionally it is possible to relax the strict rules of the machine by technical means. This is achieved by introducing buffer zones, or otherwise giving room for individual manœuvre, thus increasing the permissibility of error and individual decisions.[4] In most real situations a combination of these methods is used, but the type of combination is a value laden political decision and critically determines the way people live and enjoy or endure their work. Co-determination, room for voluntary cooperation, scope for skills, form one pole; strict discipline and perfect conformity with total management control form the other.

The questions sketched above, the extent to which the factory worker must be enslaved or may be a freely cooperating individual, clearly must be at the heart of political thinking in the context of production technology. The requirements of efficiency are a severe constraint upon the room for manœuvre in the competitive production of goods, but they should be merely regarded as a boundary condition and not a determinant. Efficiency can be safeguarded, and perhaps enhanced, in a free community of workers. The whip is not the only tool by which to achieve competitive advantage.

If a degree of self-determination and humanisation of working conditions must be a goal for every humanitarian thinker on factory production, what of the products themselves? Is it possible to find new criteria going beyond marketability, can real meaning be given to the phrase 'socially useful products' or is it no more than a statement of being against sin?

The best known attempt by workers to persuade their management to produce socially useful products in place of defence equipment is the initiative taken by the Lucas Aerospace Combined Shop Steward's Committee.[5] Lucas Aerospace is a high technology firm supplying control systems for the aerospace industry and therefore highly dependent upon military procurement. At a time of defence expenditure cutbacks the firm was faced with the prospect of considerable contraction and shedding of labour. The workers got together a committee which worked out a plan for the firm to diversify into socially useful products to replace the contracting defence market. The range of proposed products included kidney machines, heat pumps, a rail/road vehicle, a hybrid diesel/electric vehicle and other technically advanced items designed to be useful and to use the considerable skills available within the workforce. The plan absorbed and engendered a great deal of enthusiasm and cooperation but also caused major political battles with lines that were neither straight nor translucent. In the end the plan came to nought as management refused to accept the products on the grounds of their belief that no adequate markets could be found for them. The attempt to replace the lunacy of murderous defence technology by the sanity of socially useful products thus failed, but the ideal of socially useful production remained. Whether the ideal can be expressed in meaningful criteria for products of technology which might be added to, or substituted for, mere market criteria remains an open question.

Selection criteria for consumer products could fall into three categories, arranged here in descending order of acceptance:[6]

(i) market criteria;
(ii) socially responsible criteria;
(iii) additional criteria of social utility.

Market criteria clearly must be fulfilled if a firm is to produce a product in a market economy, unless public interest groups subsidise the product for one reason or another. Traditional market criteria include the competence of the firm to manufacture and market the product and are based on the hope of adequate sales and profits. These can be realised if sufficient purchasers feel that the product fulfils some of their needs at an acceptable price.

From the point of view of society as a whole, however, many additional criteria of social responsibility could be formulated. These might be grouped into (a) consumer protection, whereby society forces manufacturers to inform the purchaser correctly, to produce goods that are as safe as they can be made, and to protect the purchaser against malfunctions; and (b) prevention of waste of resources and pollution of the environment.

The first is simply an extension of the general protection of private individuals against a variety of hazards and malpractices. The judicial principle of the free market, *caveat emptor* (buyer beware), cannot suffice in a society where the buyer is confronted with the opaque mysteries of science-based machines and where any malfunction or careless design can have tragic consequences. With a plethora of new materials and new designs being offered to an unsuspecting public, the state feels more and more obliged to offer some protection against misguided purchases, inadequate technologies, or new hazards. The most obvious example might be the flammability of many man-made fabrics used for garments and for foam fillings in upholstered furniture. These are hazards against which the consumer cannot protect himself unaided.

The second category is not designed to protect the individual but rather to protect the long-term interests of society as a whole. To safeguard the future and give coming generations reasonable opportunities, it is vital that natural resources should not be squandered. Individual interests and short-term market forces cannot provide the necessary safeguards and therefore the state ought to intervene, set criteria and provide regulations and/or incentives for conservation. A very small step in this direction was made recently by requiring car manufacturers to state figures for the fuel consumption of their products. This has not only provided an added selection criterion for the consumer, but also an incentive to manufacturers to pay more attention to this aspect of their designs. Some incentives for the improved thermal insulation of houses have also been introduced in recent years and building regulations amended accordingly. A great deal more could be done and a whole arsenal of policies for conservation could be invoked, somewhat on the pattern sketched in Table 5.3. Some of these measures could provide design criteria for consumer products which would ensure sparse use of scarce materials, durability, and/or easy re-circulation of materials.

Measures for the prevention of pollution clearly are an obligation of the state and some of these measures might provide criteria for innovative products, both in the realm of consumer and of capital goods. Such criteria could include non-polluting disposal of the product after use, non-polluting use of detergents, sprays and other chemicals, reduction of smoke and toxic emissions in manufacturing processes. Regulations and incentives in these matters have to be supported by research effort and constantly revised in the light of new findings and new political will.

If these matters seem controversial, they are quite tame compared to issues we have called additional criteria of social utility. These might include concerns such as: equality of opportunity, non-interference with other consumers, alleviation of suffering, improvement of opportunities for underprivileged groups, increased cooperation between people, enhancement of creativity, satisfying employment opportunities, improved security of persons and property, improvement of health.

These concerns are not only ambitious but also value-laden and controversial. While the alleviation of suffering and improvement of health are universally applauded aims, agreement on how technical developments fostering such aims should be financed or what proportion of national effort they should occupy is anything but universal. Aims such as equality of opportunity are altogether so elusive that it must be doubted whether any technology policy can directly foster them. The assessment of impacts of technology is uncertain, and elusive aims cannot be more than added criteria in a general assessment of an innovation. None the less a systematic bias in support of policies toward politically agreed aims could act as a kind of Darwinian selection process and over the course of many years could bring the achievement of these aims distinctly closer.

Perhaps more important than positive selection might be negative discrimination. If products such as motor cars, with their tremendous social problems, were subjected to controls to curb their excesses (e.g. urban congestion, road casualties, energy guzzling, pollution, social discrimination), while competing technologies of public transport were given some support, then gradually the motor car might retreat into its spheres of greatest advantage and the withering of public transport could be halted.

Thus the set of social aims described as 'additional criteria of social utility', is seen to be only inadequately catered for by the market. If the political will existed that they should prosper, market forces would need some reinforcement by policy measures. These could provide the bias which would cause a preferential movement in one rather than the other direction. A slope of political will superimposed upon the natural landscape of the market; strong enough to cause the implementation of some social aims yet weak enough not to interfere with the freedom so essential for a dynamic society and so highly prized by the individual.

Underprivileged groups and their needs form a very special category. If sufficient effort is put into the development of technologies to aid the handicapped, or cure the sufferers from rare diseases, or give learning opportunities to underprivileged children, perhaps a considerable reduction in the totality of human suffering could be achieved. Yet the very nature of the affected groups as small minorities makes the market almost powerless to help them and all effort must come through public action. Questions of political priorities become highly controversial when competition arises between the real needs of a small group and the fulfilment of perhaps frivolous desires of a large population, especially if the frivolous desires determine the economic success of the society and thereby ultimately affect the interests of the minority groups too. The goose might lay frivolous golden eggs, but the odd one of them might serve a truly human purpose. It is tempting to establish a hierarchy of material needs— essential at one end of the scale and frivolous at the other—but any support measures with a bias towards one end must be hedged in with caution in order not to destroy the goose which lays the golden eggs and, infinitely more important, in order not to harm that precious fledgling called freedom.

SOCIAL TASKS FOR INDUSTRY

The main social task of industry is the production of products which, ultimately, satisfy some human need. Whether these be essential or frivolous, socially positive or harmful, they are all needs which the political and economic system sanctions and allows to be expressed as purchasing power. In this sense industry provides for the material needs of individuals and of society at large.

Industry does, however, fulfil at least four secondary social tasks: to provide employment, to provide a focus for other economic activity, to keep a healthy balance of payments, and to fulfil the autarkic and power aspirations of the state.

Although in most developed countries direct industrial employment now represents only about one-third of total employment, this is clearly of dominant importance. Much service employment is also related to industrial production, including transport, legal and banking services, specialist advisory and design services, research, training and education. Of even greater significance is the fact that the industrial system is indeed a system with an intricate web of mutual supplies. Thus the manufacture of one item creates opportunities for the production of many other items and, as a corollary, any decline of manufacture has far-reaching consequences in decreasing economic opportunities.

The provision of employment is a vital function. Despite the fact that the condemnation of Man to work for his living is regarded as divine punishment and Eden is seen as a place of perpetual idleness, work has become an essential cement of society. People's most important link to the society they live in consists of their economic contribution to it. Take away that link and people feel adrift, cast away and useless. For most people it is virtually impossible to carve out a meaningful and satisfactory life without playing some economic role, preferably with formal recognition in the formal economy. It is hard not to be recognised as a useful member of society.[7]

If employment provides the link to society for a great many people, the more important it is to make work itself as meaningful and as cooperative as possible. This is probably best achieved by allowing people scope to exercise their skills and their initiative with a degree of self-determination.

The fact that the industrial system functions as a system presents the greatest obstacle to industrial development. A single industrial enterprise cannot function properly without an edifice of suppliers, transportation and other infrastructural support and, most important of all, skilled and experienced workers. To raise all these out of the ground quickly is well-nigh impossible and is made even harder by shortages of capital. Hence Schumacher's concept of appropriate technology, suggesting step-wise development with optimal use of local resources.[8]

Not all countries suffer from shortages of capital and indeed the non-productive accumulation of capital in some places is one of the maladies of the world economy. For a country with large surpluses on the balance of payments it might make sense to import all or most manufactured goods. For most countries this is not a feasible procedure because they cannot afford to forgo the economic and social stimulus provided by industry and, last not least, because they could not balance their external trade without the benefit of domestic production for their own consumption and for export. Although short-term imbalances in foreign trade do not matter, a long-term imbalance leads to disaster and, generally, industry plays a crucial role in balancing trade.

The final and most elusive social task set to industry is the aim of national autarky and power. Unhappily it is only too apparent that industrial and military might have become so closely linked as to form two sides of the same coin and the whole grand enterprise called technology has become tarnished with the vision of murderous destruction. The quest for power is one of the motivations for many governments for the support of science, technology and industry. Those who think that public money could be better employed in the support of economically and socially valuable technology, rather than sophisticated means of annihilation, are told not only that defence is every nation's first priority, but also that defence technology yields valuable civilian innovations. This incidental fall-out of civilian technology from military effort is often quoted as an important justification for military R & D. In a closely knit scientific-industrial system it is, of course, inevitable that some benefit should accrue from almost any development work. With such a multitude of cross-linkages in the system, the common pool of knowledge will eventually gain even from initially secret research and useful technology may be derived from even the most extravagant defence R & D. The famous example of ceramic oven-ware as a by-product of space technology illustrates the point. It also illustrates the infinitesimally small cost effectiveness of development of civilian technology as a by-product of military requirements. It is generally more useful to aim for what you wish to achieve rather than hope that other goals will be achieved as accidental by-products, even if such by-products are often a welcome bonus and are sometimes more useful than the original aim. Those who wish to speak French should learn French, rather than take a course in Italian in the hope that the latter will help them to understand the former.

The hope of achieving military and political power through industrial might is thus an added incentive for public support of scientific research and of industry. With aspirations of power go aspirations of autarky which go beyond the economically motivated desire to build a strong industrial system and maintain a healthy balance of payments. There is a strong will to be independent of foreign suppliers in as many as possible of those industries perceived, rightly or wrongly, as basic. This will is motivated by political thinking in terms of both peace-time international competition and of possible armed conflict. Considerations of national pride provide added motivation for high technology ventures, for achievements in high technology seem to rank in prestige second only to those in sport. Unhappily, any economic activity motivated by non-economic and non-creative desires tends to weaken rather than strengthen the economic system.

NOTES

1. Winner, L., *Autonomous Technology*, Cambridge, Mass., MIT Press, 1977, p. 238.
2. Galbraith, J. K., *The New Industrial State*, Harmondsworth, Penguin Books, 2nd edn, 1974.
3. Hirsch, F., *Social Limits to Growth*, London, Routledge & Kegan Paul, 1977.

4. Gyllenhammar, P., *People at Work*, Reading, Mass., Addison Wesley, 1977; Blackler, F. H. M. and Brown, C. A., *Job Redesign and Management Control: Studies in British Leyland and Volvo*, Farnborough, Hants., Saxon House, 1978.

5. Wainwright, H. and Elliott, D., *The Lucas Plan*, London, Allison & Busby, 1982.

6. Braun, E., 'Social priorities for new technology', paper given at conference held at University of Aston in Birmingham in January 1979.

7. Jahoda, M., *Employment and Unemployment*, Cambridge, Cambridge University Press, 1982.

8. Schumacher, E. F., *Small is Beautiful*, London, Abacus, 1974; Braun, E., 'Electronics and industrial development', *Bulletin of the Institute of Development Studies*, 13 (1982), pp. 19–23.

FORCES WHICH SHAPE TECHNOLOGY

The developmental trajectory of a complex system such as technology is determined by many forces, both those of internal logic of the subject and social forces external to it. The internal logic includes both the natural development of science, where one discovery may open the path to another or one instrument may enable many new discoveries to be made, and the internal connections within technology. These include the systemic links whereby one new apparatus requires many supplementary items to facilitate its manufacture, use, maintenance and improvement. Other forces of internal logic lead to constant improvements of a technology until natural limits of performance are approached and to diversification of design. The former is the hallmark of maturing technology while the latter is more prevalent in initial attempts to carve out a niche in a new market.

The four main internal forces affecting product development are thus (i) scientific possibilities and new product ideas; (ii) systemic connections whereby any one new technology requires a range of additional developments to make its implementation practicable and efficient; (iii) the drive toward perfection; (iv) rival designs in attempts to emulate or better a new product.

The forces which ultimately lie behind all four basic manifestations are the creative will of the engineer and the corporate power of the scientific-technical establishment. The engineer wishes to invent and design new products and wishes to improve products up to the limits set by laws of nature or available materials and techniques. He wants to do these things for much the same reason as the mountaineer wants to climb Everest: to see whether he can conquer and win if he pitches his will against Nature. The scientific-engineering establishment wields sufficient power within industrial organisations to exert some pressure for the continuation of scientific-technical developments, sometimes even in cases where caution might be the better part of valour. While the accountant tends to err on the side of caution and might lead the firm to ruin through stagnation, the engineer may lead the way to ruin through excessive technical developments. In a well-managed firm a balance is held between these forces.

The primary function of technology is the development of methods of production for new and old products. The new product can only reach the market if it can be produced with adequate efficiency, while production methods for older products are subject to continuous improvement. One of the hallmarks of technology is indeed increasing productivity in the manufacture of all goods. By the twin approach of increasing productivity and overcoming barriers to mass production, technology has systematically increased the quantity and range of products and reduced their price. Technically, the key to these trends is, on the one hand, increasing refinement of production systems—in the sense of the interlinking functions leading from design and raw materials to the dispatch of finished products—and on the other hand increasing automation of production machinery. Machines become more self-acting, i.e. they work with less and less human intervention as they master increasingly complex tasks under the control of pre-programmed logic. This long-established trend has recently received a new fillip as micro-electronic devices replaced the logic provided by mechanical linkages.

The rationalisation of production through division and specialisation of labour has additionally become rationalisation through the elimination of labour. The remaining tasks are research, development and design; administration and organisation; filling of automation gaps; maintenance; machine minding; quality control; and ancillary functions such as cleaning, catering and transport. As the factor of production labour came to be replaced by the factor capital so, paradoxically, each worker became more important because he had to keep more and more machinery turning. One worker more or less in a gang of navvies does not matter, but every worker in a highly automated factory is crucial.

Gradually wages rose. The worker became more important and politically better organised, thus the pressure for higher wages became more effective. Increasing substitution of capital for labour and rising labour productivity made it economically possible to pay higher wages and, finally, the need to sell an increasing amount of goods made it necessary to provide the worker with more purchasing power. The rising cost of labour became a feature of mature technological society and gave a further twist to the automation spiral— a kind of positive feedback in the economic system favouring increasing automation in the technical system. The social repercussions of increasing costs of labour were enormous and cannot be fully treated here. Let it suffice to mention the virtual elimination of domestic labour with its concomitant trend to labour-saving household goods and machinery; and the importation by many developed countries of labour from poorer countries to fill vacant low-paid labouring jobs and overcome bottlenecks in economic growth. Both these factors have far-reaching consequences which have changed the social and political face of several European countries.

The other dominant feature of mature technological society—the feature we took as the mark distinguishing it from industrial society —is the enormously increased systematic use of science. Technology has its roots in the crafts and the early engineers were craftsmen—blacksmiths, millwrights, instrument makers. Science on the other hand started from philosophy, from the eternal question of how the world was constituted. The two strands began to meet with the increasing scientific use of craft-made instruments, the establishment of the experimental/observational method in science and early practical scientific applications in material science, navigation, clock-making, weaponry and medicine. These developments are, roughly speaking, characteristic of the seventeenth and early eighteenth centuries. The first major industrial applications of science came with the establishment of the chemical industry, followed by electrical engineering. From that time the development of engineering and science became increasingly intertwined and the training of engineers and other practitioners became increasingly scientific. The second half of the nineteenth century saw an increasing recognition of the importance of science and technology by the state. The period is characterised by the great exhibitions, the establishment of what are now called technical universities and of government research labora-tories and standards institutions. The early twentieth century saw the increasing establishment of organised scientific research and the Second World War put the final seal on the modern systemic links between science—technology and industry, with a support system of universities and research institutes.

Science and technology have become associated, allied and inter-linked to the point of near indistinguishability. The total research system has grown enormously and now provides an almost autono-mous economic force which drives technology on a path of accelerat-ing innovation. Unfortunately a relentless law of diminishing returns operates which demands greater and greater R & D efforts for smaller and smaller improvements. As a technology matures, this law takes a heavy toll. The R & D establishment nevertheless pushes efforts to approach the natural limits of a technology as closely as possible. Sometimes the way out of the impasse is provided by the appearance of a superior rival technology which by-passes the particular bottleneck reached by the previous technique. The creativity and drive to achievement, the will to climb the highest peak and competition within the peer group drive the engineer and through him technology is spurred on.

Competition among engineers and scientists is only one manifestation of a fundamental driving force for technology. Competition probably lies at the root of more of the multifarious forces which determine the course of technology than any other root cause and figures conspicuously among the social forces both internal and external to the industrial–scientific–technical system.

Within the industrial system, competition for market shares and for profits can be severe. The results are pressures toward certain kinds of product innovation and product differentiation, aggressive marketing, and a constant drive toward greater factor productivity. A climate in which efficiency is the ultimate arbiter ensues, which Ellul calls 'technique' and condemns wholesale. Competition in the capitalist market economy is not, however, the only force favouring efficiency. The drive of the engineer toward the limits of the possible; the desire of the manager to control the work-force and the work-process; the desire of people for the cheap and plentiful supply of goods are but three further causes for the quest for efficiency. It might be noted that some at least of these forces are independent of the capitalist market economy. As long as the main purpose of industry is the supply of goods, efficiency is bound to be pursued as an aim. Efficiency must, however, be tempered with human understanding and a degree of slack if it is to be compatible with tolerable social conditions.

Among the external driving forces of technology pride of place goes to some quite fundamental human desires. The biblical curse of earning one's daily bread in the sweat of one's brow must be taken in juxtaposition with the dream of paradise in which sustenance is achieved without effort. Technology is to ease the burden of performing daily tasks and thus rid Man of the ancient curse. But beyond the easing of the burden of necessary tasks, technology can provide comfort, pleasure, beauty and worldly possessions. It can be the expression of desires such as acquisitiveness and playfulness; the fulfilment of the urge of achievement and creativity and can satisfy a sense of curiosity and wonder.

Competition in the personal sphere can also involve technology. The entrepreneur seeking to make a success of his firm does so not only because of his desire for riches, but also because he wishes to win in the game and get ahead of the pack. He wants high profits, but even more he desires a respected position, an association with a firm of high repute, and technical excellence and elegance. This latter desire by engineers and entrepreneurs is one of the checks on market forces which drive technology toward the lowest common denominator of mediocrity.

In the private sphere people compete for position in society and this competition takes many forms. One is the acquisition of large quantities of products, particularly high prestige products and positional goods. Although inherently products of technology cannot have positional value as there is no uniqueness associated with them, all manner of device is used to circumvent this basic rule. Some products are made to very high standards at a very high price; others are 'personalised' by various additions and special order features; yet others are issued in deliberately limited editions. Apart from all this, prestige is deemed to be associated with early ownership of a new product—a fact which eases the premature marketing of goods and thus can cause the proud owner many a headache.

Among the fundamental human desires which technology can and does cater for is, as mentioned several times before, the desire for personal autarky. With technology acting as an amplifier of personal capabilities it enables people to achieve single-handed what previously required human help and cooperation. As technology also provides many requirements, such as stored food supplies, heat, water, light, waste disposal, washing and cleaning, at the flick of a switch and in apparent independence from other humans, the individual can achieve a status of an illusory self-sufficiency. The fact that this power of technology satisfies only one pole of the twin desire of autarky and cooperation is one of the causes of modern Man's isolation.

The necessarily incomplete summary of an analysis of forces which shape technology needs to be supplemented by another fundamental human requirement; that of security. Throughout history people were prepared or forced to forgo some freedom and self-determination and private consumption in exchange for protection against internal and external enemies. In recent years the loss of personal consumption has become very large, as a high proportion of total available research effort and industrial production have been absorbed in vast military machines. This is as true, though more tragic, in many developing countries as it is in many developed ones. Because of the large R & D effort going into military goods and the non-market conditions under which these goods are procured, severe distortions of the technological pathway occur in addition to civilian loss of consumption. Unhappily, with all the huge price paid for security by so many nations, the world appears to be as dangerous, or more dangerous a place as it ever has been. It would hardly be fair to blame technology for this state of affairs, despite all her willingness to be the hand-maiden of power politics.

POST INDUSTRIAL SOCIETY

The hallmark of modern technology is efficiency of production—a high ratio of outputs to inputs—or high productivity. Until very recently this has been hailed as a major triumph and the rate of increase of productivity was taken as a measure of economic success and a portent of future good fortune, for growing productivity is a cause of increasing wealth. The slowing down of productivity growth in the early seventies is regarded as one of the signals of the economic crisis which has been deepening ever since.

With the arrival of microelectronics and computer aided enhancement in the feasible efficiency of production and administration, the argument has apparently been turned round and increasing productivity is now thought to bring the spectre of unemployment in its wake. This is curious, as rising productivity should inherently raise the wealth of any nation. Unemployment can result from it only if consumption, for one reason or another, does not rise to meet increased potential production. This can happen for a number of reasons, such as saturated markets for old products and insufficient innovation, lack of purchasing power, structural weaknesses, or any other imbalances or rigidities in the national or international economic system. Productivity increases lead to economic growth only if the benefits obtained from them are distributed from their point of origin to permeate the economy and lead to increased effective demand. If markets for certain goods are saturated, then additional demands can be stimulated and satisfied. Among such additional demands may well be those for private or public services and even demands for additional leisure.

Because of the apparent stagnation of markets for many industrial products and the manifold difficulties of adding more and more products to those already on offer in order to maintain consumption in the face of rising productivity, many authors have considered that the consumption of services would gain increasing significance. Among the many difficulties of maintaining rising material consumption are problems of finding suitable innovations, consumer resistance to the notion that happiness equals the purchase of industrial products, and problems of exhaustion of natural resources. Services, on the other hand, suffer fewer problems of saturation and meet many aspirations of a good life and a fair, caring society.

The term 'service society' has been used to describe a society in which industrial activities would be secondary to the provision of consumer services.[1] There are some wild estimates that 10 per cent of the working population will be sufficient to produce all the goods society may require and everybody else will either be engaged on providing services or will enjoy creative leisure. Such estimates seem highly unlikely. They are based on the tiny proportion of workers remaining in agriculture and on the steady reduction in the proportion of industrial workers which took place in recent years. Against this it must be remembered that agricultural productivity is backed by large industrial inputs, that agriculture produces a tiny assortment of simple products compared to the almost infinite variety of industrial production and that many service activities are directly related to industry and could properly be considered part of it. We ought to remember also that most services depend on industrial inputs: no transport without major technical provision, no leisure activities without massive technological underpinning, no banking without safes, computers, telecommunications. If we consider that building, construction, and the massive urban infrastructure are all products of industrial technology, then any massive reductions in the proportion of industrial workers appears unlikely even in the face of all foreseeable productivity gains. The proportion may further decrease, depending on what is counted as industry, but the decrease is likely to be small and slow.

Against the concept of a service society a kind of self-service society was postulated by Gershuny.[2] The argument is roughly that because of the great productivity of industry and the high cost of labour, and therefore of labour-intensive services, consumers rely more and more on self-service supported by industrial products. The domestic washing machine and other domestic equipment replace the laundry and domestic services and, to some extent, restaurant services. Similarly a whole range of 'do-it-yourself' equipment replaces the plumber, decorator, electrician and carpenter. Garden machinery takes the place of the gardener, many car owners service their vehicles themselves.

It would seem that Gershuny's argument is in accord with our thesis which asserts that technology is best able to satisfy autarkic needs. The self-service society is one with a high degree of autarky; each person is apparently independent from other persons, although highly dependent upon a functioning system of industrial production and infrastructure. The dilemma of the one-sided capability of technology to support autarkic needs is that cooperative needs are neglected. The very efficiency of technology drives people into their self-sufficient little castles and undermines cooperation and the provision of services based on people doing something for each other. The division of labour in the service sector too becomes a displacement of labour by machines.

The trend is not uniform. While in Britain personal services have atrophied rather badly, in other European countries people do spend a high proportion of their rising incomes on eating out, holidays, art and entertainment and even craftsmen are still generally available. Perhaps it is no accident that Gershuny's book was written in Britain, while Bell's home is in the United States, where the concept of service has not been entirely displaced by that of self-service. But be this as it may, it seems unlikely that future society will consist of a mix of services and industry all that different from the current mix. In particular if we distinguish between services as items of final consumption and intermediate services, the recent move toward more service occupations would appear less marked.

Slogans describing society abound. Not only have a post-industrial service society and an after-industrial self-service society been postulated, but we are also supposed to be or become an information society. When all is said and done, all human activity is and always has been based on information—indeed any action other than random motion is informed. With increased complexity of society and increased division of labour the acquisition, storage, transmission and manipulation of information has become more formally articulated and more divorced from physical activities. Whereas previously people simply knew how to do things and transformed their knowledge into action, now the knowledge leading to action is often handled separately, as is almost symbolically shown in the extreme example of a numerically controlled machine tool where the knowledge is carried in explicit language on tape and the translation into action is performed by the machine. As economic and social units have become larger, so more coordinating and controlling procedures had to be introduced and a vast administrative apparatus grew up. It is the separation of knowing from doing and the need for administration which gave rise to the concept of information handling as a separate and increasingly dominant activity. Porat estimated that in 1967, 46 per cent of all economic activity in the United States was primarily concerned with information.[3] Whether this estimate is accurate or even meaningful is immaterial; the important current concern is that the major activity of information handling has become not only separated from other activities but has acquired its own technology. We see on the one hand a vast growth in formal institutions whose task is the creation of knowledge—the much publicized institutionalisation of research—and on the other hand the creation of a new class of technology, so aptly called *informatique* (informatics), which stores, manipulates and transmits information of every kind.

A vast body of speculative and empirical research has, as might be expected, made it its business to estimate the impact of the new technology upon society; the way we live and above all, on the number and kinds of jobs which will be available. To say that the studies are inconclusive is merely stating the obvious; but perhaps one conclusion may be permitted. The universal law of diminishing returns, which dominates so many human activities, operates with a vengeance as far as increased efficiency of information processing is concerned. For information can be good or bad, useful or useless, and no matter how well machines process it, ultimately it has to be both created and critically absorbed by humans. Trees do not grow up to the sky and neither will information technology. Future society, whatever we might call it, will hopefully still depend on human judgement and will still contain an enormous variety of tasks and occupations.

The trend towards formalisation of information and growth in informatics will no doubt continue but any gains in efficiency thus obtained will only be beneficial if new needs can be satisfied. One such need might be leisure and no doubt the trend toward shorter working weeks, years or lives will also continue. At present this is seen as an escape from the impasse of stagnating effective demand lagging behind enhanced productivity and thus giving rise to unemployment. In better times any reduction in the commitment to work is seen as a worthwhile aim and regarded as consumption of leisure.

Undoubtedly leisure is a valuable commodity, provided people have sufficient money and sufficient facilities to use it well. Very few aspire to the ideal of the idle poor. How far the commitment to work can be reduced without at the same time removing from people their link to society, their self-esteem as valuable members of their social group, is a matter of conjecture. The unemployed certainly suffer badly from feelings of being unwanted and unnecessary and from difficulties of structuring their time, quite apart from not having money or facilities to enjoy their unwanted leisure. Although more leisure will probably be a hallmark of the future, hopefully the majority of the population will participate in the wealth creating process and thus feel as legitimate members of society.

There probably are natural limits to the consumption of industrially produced material goods. These are not reached, particularly because of maldistributions of incomes and wealth which leave large sections of the population in even the most developed countries devoid of the means to purchase their fair share of industrial production. Nevertheless there are limits to consumption set on the one hand by the capacity of nature to provide energy and raw materials and to absorb industrial waste, including excess heat, and on the other hand by the mutual interference of consumption, especially conspicuous in transport. Thus it is desirable and likely that future economic growth should occur to a considerable extent in services as items of final consumption, whether these be educational, recreational, health care or any other. As it is difficult for some of these and the many infrastructural services of a modern state to be either adequately marketed by private finance or publicly financed through taxation, some new ways of financing various services will have to be found. This might consist of a mix of self-financing marketable services and state finance raised in novel ways, perhaps not by removing money from people's pockets but more painlessly by not putting it into their pockets in the first place.

All the discussions about saturating markets for industrial products and resultant dangers of unemployment must seem cynically comical to the urban poor of many developed countries and to the millions of poor people in Third World countries. The problem of urban poverty is exacerbated by lack of attractive investment opportunities in urban renewal. The decaying buildings, rotting sewers and derelict land of many a city could be reinstated if suitable returns on invested capital were in prospect for private investors. As this is hard to achieve, especially in the case of public utilities such as sewers, city or state authorities have to make the investment and then all the problems of public financing are compounded by problems of setting priorities. It is hard to escape the conclusion that public investment ought to be possible in lieu of the appalling waste of unemployment.

Problems of poverty in developing countries are too vast and too complex to be even touched upon in our context. The only thing that may be said is that industrial development might provide a partial answer, but such development is exceedingly difficult because industry operates as an interdependent system of many separate parts and is also dependent upon a complex infrastructure. Such a system cannot grow quickly but takes time and time is at a premium while populations grow and the developed world does not stand still. There is no simple recipe, no one road to salvation. Every means must be used, including technology transfer, foreign aid, foreign investment, improvement in the terms of trade and building up of an infrastructure, including trained people at all levels. Every scrap of competitive advantage needs to be explored with local initiative and used to the full. Although the term appropriate technology may be but a tautology, the question of what is appropriate ought to be at the forefront of everybody's mind.

TOWARD A POLICY FOR TECHNOLOGY

Technology policy is carried out, consciously or unconsciously, not only by many government departments but also by decision makers in industry and even by private purchasers of products of technology. Our discussion will concentrate on desirable aims for technology policy and will thus be concerned with all these groups, although of course government departments are the most effectively conscious carriers of any societal policy aims.

Undoubtedly the total system of society and its technology is too large and too complex to be either completely understood or effectively steered. Nevertheless partial understanding is valuable and even small corrections to the course events would take without conscious policy interventions may, in the longer run, have considerable effects. Technology is tossed and buffeted by many forces—internal and social—and as a result is a wayward creature of human ingenuity. Its unpredictable and self-willed nature gives technology an appearance of autonomy, yet it is of society and bound with it by a thousand bonds. Wayward creature that technology may be, its course set by a multiplicity of forces and influences, it is yet amenable to conscious social policy measures as these constitute one of the many influences. The aim of policy is not perhaps to tame the beast, but to curb its excesses and guide its power to where society most needs it.

The course of technology often seems inevitable, yet the logic of history is only a result of particular, almost accidental, constellations of forces and the apparent inevitability is the cosy logic of hindsight. The course of technology and its social repercussions are complex and not fully foreseeable. Technological innovation and technology policy are uncertain ventures guided only by partial, tentative knowledge. Perhaps this is characteristic of all human endeavour and certainty of knowledge is only given in exchange for simplicity of mind. Yet the improvement of knowledge and greater efficacy of policy must remain important goals for thought and for research.

The question of the aims and scope of technology policy is the central question of our discussion and the time has come to tackle it explicitly, albeit summarily.

The first aim of public technology policy must be to compensate for weaknesses of the market. There are many, and without any claim to completeness we shall discuss some of the more obvious ones.

Almost by definition, the market is unable to give adequate representation to the needs of underprivileged groups. Whether they be socially underprivileged and thus of limited purchasing power and limited political representation, or naturally underprivileged by some physical or mental handicap, the ruthless rules of survival in a market economy often leave such groups without the technical support which might be feasible on purely technical grounds. Thus the support of research and development, and even the purchase, of technical aids to the handicapped, of cures for rare diseases, teaching aids for the undereducated are all aspects of the same political willingness or unwillingness to help those who cannot help themselves in a society geared to the satisfaction of the needs of the many rather than the few.

The second weakness of the market is its inability to take the long-term view. By its very nature the market caters for the here and now, is a snapshot of forces, and disregards the future. Coupled with the inability to consider the future, it must be very seriously doubted whether the attempted maximisation of private benefits represented by the market truly achieves the optimum in public benefit. Adam Smith's invisible hand can be pretty cruel and needs to be guided with a light touch of the visible hand of government. Long-term concerns of conservation of natural resources, of preservation of the natural environment and of urban planning are obvious cases in point. There are many other, less obvious, examples which should be objectives of policy. One area in particular stands out; the technology and organisation of production. Instead of the attempt to automate all production completely and using humans only to fill in gaps in automation or to shore up temporary deficiencies caused by technical barriers to automation, human requirements and capabilities ought to be important parameters in the organisation of production. Of course factories live by their products and their ultimate aim is the production of competitively marketable goods. Yet factories also give employment to people and are the place of one of their main activities. Disregarding human needs is cruel, while disregarding human abilities is wasteful. Competitive forces, allied with technical ingenuity, are driving production technology along a path where humans might become but unwanted cogs in otherwise technically perfect production machinery. Competitive forces and technical ingenuity need to be modified and guided by consideration of human needs and aspirations not only in the market place but also in the place of work, the factory. The preservation of skills (not necessarily the old crafts but skill in a more general sense, the use of human abilities of judgement and decision) should be aims of policy. Getting away from inexorable pacing by machine, giving room for manoeuvre and initiative, a breathing space for life among the deadly monotony of mechanical operations, should be achievable with very small losses in efficiency and very large gains in happiness. Adding the worker's mental well-being as a legitimate concern to his physical safety and bodily health would undoubtedly pay dividends in all round health of society.

Objectives of this nature are elusive and intangible and militate against much traditional managerial and political thinking. Yet in constructive discussion between workers, engineers, managers and politicians it should be possible to advance beyond the stage of vague humanitarian aims into the arena of real policy measures. Programmes of humanisation of working life, legislation on safety and health, the famous Volvo experiment and a miscellany of technology agreements[4] have set first markers in this new, promising and long-overdue endeavour. Certainly the few steps taken so far in this direction have shown that the thought of considering the industrial worker as a human being rather than an extension to a machine is neither Utopian nor economically prohibitive.

One of the requirements of a humane production technology is the strengthening of cooperation between individual workers. There can be no universal prescription on how to do this, but certainly the division of labour allows for cooperation between those participating in the whole process and does not require that all coordination of effort should be removed from the participants and placed solely in the care of a separate coordinating function or authority. The loss of cooperation on the factory floor is but one aspect of the general trend of technology to aid the isolation and autarky of people rather than their cooperation. While some amelioration of this tendency at the production level is possible by organisational and technical means, there is little that technology can do at the broader social level. Compensation for the tendency of technology to drive people along the path to autarky and away from cooperation needs to be achieved by socio-political measures.

It must be recognised that one of the consequences of technical advances is that many people have become isolated in their work and in their private life. There are a few technical remedies to this at work level, but essentially technological autarky must be balanced by social cooperation. Any policy measures and initiatives which strengthen cooperation must be welcomed, provided they do not attempt to achieve social cohesion by uniting people against other groups. Unhappily the most readily plausible common causes are those directed against outside groups, yet that sort of unity does not increase the sum total of cooperation, it merely polarises it. On the other hand all forms of constructive cooperation at enterprise or community or any other level seem highly desirable. Forms of cooperative ownership, sports and social clubs, community projects of all kinds, spare time educational and cultural activities, all these are suitable to counteract the isolating effect of technology.

Any technology policy within a market economy should define certain characteristics it is seeking to foster and use these as selection criteria for its support of research and of innovation. Naturally the criteria can never form a closed group—they must be open to constant revision and must not exclude specially significant new ideas which perhaps fall somewhat outside the chosen characteristics. Any policy must be viewed as a market corrective, not as its replacement.

A socially useful technology would have several attributes in various mixes. Among the social purposes which might guide technology policy, we would include the following: conservation of natural resources; protection of the natural environment; improvement of the urban environment; safety; creative use of skills; promotion of health; promotion of cooperation; provision of creative leisure facilities; help to underprivileged groups; lack of interference with other users.

Good as these attributes may be, it must always be remembered that technology functions as a system and many techniques are needed to produce any one product. Technology also provides all the basic requirements of society, including a vital infrastructure. The point of an enlightened technology policy is that it must balance the competitive forces in a market economy and aid the higher and more long term aspirations of society. Technology must not, of course, be hemmed in too much as otherwise imagination will not flourish and both internal and external markets will be unable to develop for lack of technological stimulation. All policy measures must aim to push technology in desirable directions but must not hinder developments unless these are positively harmful.

Thus the reverse side of promotional technology policy must be regulatory and restrictive in order to prevent excesses and dangers to health, nature or social well-being.

The basis for promotion of technology as well as for curbs upon it is provided by an assessment of the potential effects of the particular technology. Thus some form of Technology Assessment is indispensible, despite all the difficulties with which the practical execution of Technology Assessments is fraught. Neither promotion nor prohibition are possible without assessment, but it ought to be remembered that the best weapon against uncertainty is flexibility and that each assessment requires frequent reappraisal.

When it is decided to promote an innovation, perhaps because the market cannot support it or because the risk or effort involved are too great to be carried by private enterprise, or because of international competition, then the support measures must be finely tuned to the specific situation and its requirements. While pure research, though probably with some awareness of possible applications, may generally aid innovation in a particular field, it will do little or nothing to help a small firm launch a new product. If that is desired, support may be necessary not just during development phases of the product but right up to initial production and marketing. The donor of support for technological innovation needs many hands—as the god Shiva—to give in different ways for different purposes.

International competition is one of the forces which drive firms into innovation and governments into support policies. Indeed a new type of competition has arisen whereby countries vie with each other in attracting industrial investment by various support measures and in effect also compete with each other in innovation support. The gunboat and tariffs have given way to technology support programmes. Although the new method is infinitely preferable to the old, it is still questionable whether there is any overall benefit from this zero-sum game. Perhaps one day trade agreements will cover this aspect of international competition.

Among the elemental forces driving technology, competition takes pride of place. Many of the basic industries, especially those who produce undifferentiated standard commodities like steel or bulk chemicals, compete world-wide on price alone and this is proving truly murderous with present over-capacity. All firms attempt to reduce their production costs by a combination of process innovation and organisational rationalisation, but even so such competition leaves little or no scope for adequate profits. In fact the losses of many steel-makers have reached alarming proportions and have caused major job losses in the industry and dependent economic activities. No doubt world over-capacity in industries producing undifferentiated basic commodities is both a cause and a result of recession. It is to be expected that in the future these industries will reach new equilibrium levels of production and that world trade in manufactured basic commodities will be relatively small as many countries will be self-sufficient in this respect.

On the other hand competition in products where differentiation is possible depends not on price alone, but on a combination of price, design, quality, delivery time, service and indefinable consumer appeal. The competitive erosion of prices of established products forces manufacturers into ever new products or product improvements, even if these occasionally consist only of cosmetic re-designs.

Competition may be a Darwinian struggle for survival. In a market economy, individual firms are forced, by and large, to produce their wares at a price comparable to that of the competition or to produce superior goods. Unless they have special advantages, such as tariff protection or subsidy or low wage rates, they are forced, in established products, to use comparable production techniques and produce comparable goods. There is, of course, some important scope for individual corporate strategies and variations, but the need to survive in a competitive world imposes considerable constraints. As the first imperative of action for most organisations and all organisms is to ensure continued existence, the threat of being extinguished by competitors with superior technology is a powerful one. If firms are forced by competition to adopt the most efficient production technology, they are also forced to produce goods of competitive design and quality. Thus innovation in one firm forces innovation in another and some firms will innovate aggressively in order to get just a little ahead of the pack. In the capitalist system, particularly in the United States, there are many potential entrepreneurs trying to find niches in which a new technology might satisfy a previously undetected or unstimulated latent need or, more rarely, an articulated but unsatisfied need. In this way the entrepreneur attempts to get ahead, to reap financial rewards and the sweet feeling of success. Sadly, the very possibility of getting ahead and succeeding entails the alternative of falling behind and failing. Many do.

Although competition undoubtedly brings forth much ingenuity, many innovations and many price reductions from which the consumer and the economy draw considerable benefit, it also imposes large costs. Costs in human terms of stress, pressure, aggression and sometimes failure. Costs also in terms of subservience of people to the god efficiency and the enforcement of working conditions and pace which are inhuman in many senses. There are economic and environmental costs to excessive competition too. If machines and equipment become obsolete before they are worn out their replacement imposes a perhaps unnecessary cost. If obsolescence is built into goods, their premature replacement imposes a cost. In both cases part of the cost is a squandering of energy and raw materials which are embodied in the machines.

Although competition must certainly be regarded as a positive force, it becomes destructive if allowed too free a rein and if not compensated by considerations of humanity, the environment, cooperation and other values. Efficiency is valuable, but many other things ought to be valued more highly. The greatest efficiency is not the road to the greatest happiness. This kind of consideration has many repercussions in general political thinking and these are outside our present scope. They also have repercussions in technology policy. These might be expressed in attempts to curb international competition in support policies but might also be expressed in many other ways. Only a few possible avenues will be mentioned and recommended for further exploration.

(i) Cooperative ventures of many kinds could be fostered, starting with industrial research institutes and ending with cooperative ownership of enterprises. (ii) The competitive climate need not be unnecessarily fostered by too strict an adherence to rules about cooperation between firms. Clearly cartels and price fixing are undesirable, but so is cut-throat competition. Technical cooperation between firms is no bad thing at all. (iii) Unnecessary innovation should not be supported by the state. Of course no curbs should be put on it, but allowing something or deliberately supporting it are two very different things. All support for innovation ought to be carefully weighed on social grounds—as discussed earlier—but also on grounds of whether it fosters excessive competition. (iv) Rules and regulations and support measures on the use of energy and materials could be formulated in such ways as to curb unnecessary competition but foster its useful aspects. Genuinely new solutions could be positively supported, while excessive use of materials and energy, including that by excessive corrosion or obsolescence could be retarded. Rules about writing-off of capital equipment might be one way of doing this.

When all is said and done, technology must be allowed considerable freedom to develop in its own wayward way. Technology is so much part of society, that any curbs on its freedom could become curbs on the freedom of society. Yet the very nature of technology as an aspect of society means that it must be subject to certain rules of social intercourse, for total freedom is in the end self-destructive. Technology can be steered a little and if the steering is done with sensitivity, wisdom and knowledge it can greatly enhance the value of technology while curbing its excesses. In the final analysis technology policy is part of the body politic and therefore full of controversy, but valuable consensus can be obtained on many of its aspects and this should be actively sought. If technology be a ship on uncharted seas, tossed by many social forces and propelled by its internal logic, let technology policy provide a sensitive helmsman to guide the ship on a prosperous voyage to an unknown destination.

NOTES

1. Bell, D., *The Coming of Post-Industrial Society*, London, Heinemann, 1974; see also Haller, M., 'Auf dem Weg zur Dienstleistungsgesellschaft?', *Wirtschaft und Gesellschaft*, 8, 1982 pp. 607–54.
2. Gershuny, J. I., *After Industrial Society?*, London, Macmillan, 1978.
3. Porat, M. U., *The Information Economy*, Office of Telecommunications Special Publication 77–12, Washington, US Department of Commerce, 1977.
4. Steward, F., and Williams, R., 'Trade unions and technological change', Unit 9, *Control of Technology*, Open University, 1983 and Williams, R., and Moseley, R., 'Technology agreements: consensus control and technical change in the workplace', in Bjorn-Anderson, N., *et al., Information Society —for Richer for Poorer*, London, North-Holland, 1982, p. 231.

Index